Post-COVID Tourism - Tendencies and Management Approaches

Edited by Rui Alexandre Castanho, Mara Franco and José Manuel Naranjo Gómez

Published in London, United Kingdom

Post-COVID Tourism - Tendencies and Management Approaches
http://dx.doi.org/10.5772/intechopen.104240
Edited by Rui Alexandre Castanho, Mara Franco and José Manuel Naranjo Gómez

Contributors
Ajay Khatter, Alain A. Grenier, Alfredo Huamán Cuya, Ali Said Akaak, Artemio Montesinos Palomino, Asma Akmal, Catarina Mesquita, Charalampos Vazouras, Cláudia Mendeiros, Fangzhou Wan, Felipe Rafael Valle Díaz, Fentanew Alelegn Masrie, G. Somu, Harshvardhan Sai Sadineni, Helena Albuquerque, José Manuel Naranjo Gómez, Kazuo Nishii, Khalid Salim Al-Shanfari, Konstantinos Varvaropoulos, Mara Franco, Marília Durão, Michael Mncedisi Willie, Mnghui Yang, Mohammed Nazeer Ahmed, Navaneetha Krishnan Rajagopal, Oscar Apaza Apaza, Percy Lima Román, Rita Peres, Rosmel Iván Rodriguez Peceros, Rossana Santos, Rui Alexandre Castanho, Spyros Avdimiotis, Sónia Nogueira, Takeshi Kurihara, Tesfaye Fentaw Nigatu, Thuy Nga Phan, Yilin Sun, Yinan Dong, Yubin Xi, Zelalem Getnet Ambaw

First published in London, United Kingdom, 2024 by IntechOpen
IntechOpen is the global imprint of INTECHOPEN LIMITED, registered in England and Wales, registration number: 11086078, 167-169 Great Portland Street, London, W1W 5PF, United Kingdom

British Library Cataloguing-in-Publication Data
A catalogue record for this book is available from the British Library

Additional hard and PDF copies can be obtained from orders@intechopen.com

Post-COVID Tourism - Tendencies and Management Approaches
Edited by Rui Alexandre Castanho, Mara Franco and José Manuel Naranjo Gómez
p. cm.

This title is part of the Sustainable Development Book Series, Volume 19
Topic: Sustainable Economy and Fair Society
Series Editor: Usha Iyer-Raniga
Topic Editor: Antonella Petrillo

Print ISBN 978-1-83769-141-8
Online ISBN 978-1-83769-142-5
eBook (PDF) ISBN 978-1-83769-143-2
ISSN 2753-6580

An electronic version of this book is freely available, thanks to the partial funding support of libraries working with Knowledge Unlatched. KU is a collaborative initiative designed to make high quality books Open Access for the public good. More information about the initiative and links to the Open Access version can be found at www.knowledgeunlatched.org

For EU product safety concerns :
IN TECH d . o . o . , Prolaz Marije Krucifikse Kozulić 3 , 51000 Rijeka , Croatia ,
info@intechopen . com or visit our website at intechopen . com .

IntechOpen Book Series
Sustainable Development
Volume 19

Aims and Scope of the Series

Transforming our World: the 2030 Agenda for Sustainable Development endorsed by United Nations and 193 Member States, came into effect on Jan 1, 2016, to guide decision making and actions to the year 2030 and beyond. Central to this Agenda are 17 Goals, 169 associated targets and over 230 indicators that are reviewed annually. The vision envisaged in the implementation of the SDGs is centered on the five Ps: People, Planet, Prosperity, Peace and Partnership. This call for renewed focused efforts ensure we have a safe and healthy planet for current and future generations.

This Series focuses on covering research and applied research involving the five Ps through the following topics:

1. Sustainable Economy and Fair Society that relates to SDG 1 on No Poverty, SDG 2 on Zero Hunger, SDG 8 on Decent Work and Economic Growth, SDG 10 on Reduced Inequalities, SDG 12 on Responsible Consumption and Production, and SDG 17 Partnership for the Goals

2. Health and Wellbeing focusing on SDG 3 on Good Health and Wellbeing and SDG 6 on Clean Water and Sanitation

3. Inclusivity and Social Equality involving SDG 4 on Quality Education, SDG 5 on Gender Equality, and SDG 16 on Peace, Justice and Strong Institutions

4. Climate Change and Environmental Sustainability comprising SDG 13 on Climate Action, SDG 14 on Life Below Water, and SDG 15 on Life on Land

5. Urban Planning and Environmental Management embracing SDG 7 on Affordable Clean Energy, SDG 9 on Industry, Innovation and Infrastructure, and SDG 11 on Sustainable Cities and Communities.

The series also seeks to support the use of cross cutting SDGs, as many of the goals listed above, targets and indicators are all interconnected to impact our lives and the decisions we make on a daily basis, making them impossible to tie to a single topic.

Meet the Series Editor

Usha Iyer-Raniga is a professor in the School of Property and Construction Management at RMIT University. Usha co-leads the One Planet Network's Sustainable Buildings and Construction Programme (SBC), a United Nations 10 Year Framework of Programmes on Sustainable Consumption and Production (UN 10FYP SCP) aligned with Sustainable Development Goal 12. The work also directly impacts SDG 11 on Sustainable Cities and Communities. She completed her undergraduate degree as an architect before obtaining her Masters degree from Canada and her Doctorate in Australia. Usha has been a keynote speaker as well as an invited speaker at national and international conferences, seminars and workshops. Her teaching experience includes teaching in Asian countries. She has advised Austrade, APEC, national, state and local governments. She serves as a reviewer and a member of the scientific committee for national and international refereed journals and refereed conferences. She is on the editorial board for refereed journals and has worked on Special Issues. Usha has served and continues to serve on the Boards of several not-for-profit organisations and she has also served as panel judge for a number of awards including the Premiers Sustainability Award in Victoria and the International Green Gown Awards. Usha has published over 100 publications, including research and consulting reports. Her publications cover a wide range of scientific and technical research publications that include edited books, book chapters, refereed journals, refereed conference papers and reports for local, state and federal government clients. She has also produced podcasts for various organisations and participated in media interviews. She has received state, national and international funding worth over USD $25 million. Usha has been awarded the Quarterly Franklin Membership by London Journals Press (UK). Her biography has been included in the Marquis Who's Who in the World® 2018, 2016 (33rd Edition), along with approximately 55,000 of the most accomplished men and women from around the world, including luminaries as U.N. Secretary-General Ban Ki-moon. In 2017, Usha was awarded the Marquis Who's Who Lifetime Achiever Award.

Meet the Volume Editors

Rui Alexandre Castanho holds an International Ph.D. in sustainable planning in borderlands. Nowadays, he is a professor at the WSB University, Poland, and a visitor professor at the University of Johannesburg, South Africa. He is a European Climate Pact Ambassador. Besides, he completed post-doc research on the GREAT Project, at the University of Azores, Ponta Delgada, Portugal. Furthermore, he collaborates with the ECIDES Research Group—University International of La Rioja (UNIR), Spain; VALORIZA—Research Center for the Enhancement of Endogenous Resources, Polytechnic Institute of Portalegre (IPP), Portugal; CITUR—Madeira Centre for Tourism Research, Development and Innovation, Madeira, Portugal (vice-coordinator); and AQUAGEO Research Group, University of Campinas (UNICAMP), Brazil. Moreover, he has graduated in green spaces engineering and agricultural engineering and holds a master's degree in planning, audit, and control in urban green spaces. He is also the author and/or editor of more than 20 books, more than 140 indexed peer review papers, 90 indexed book chapters, and 100 conference papers and has made several presentations worldwide academic events as a keynote speaker.

José Manuel Naranjo Gómez holds a Ph.D. in geodesy and cartography, an engineering degree in geodesy and cartography, and an engineering degree in surveying from the University of Extremadura. He has worked as a professor for the University of Extremadura since 2002 and has research in the fields of sustainability, transport infrastructure, territorial accessibility, land uses, and territorial management. Also, he is a member of the University Research Institute for Sustainable Territorial Development (INTERRA), the coordinator of the researching group Environment and Spatial Planning, and he is a researcher on Research Center for the Enhancement of Endogenous Resources (VALORIZA-IPP).

Mara Franco is a Ph.D. in business and management studies from the School of Economics and Management of the University of Porto. She is an assistant professor lecturing curricular units in the marketing and strategy fields and is an integrated researcher of CiTUR-UMa. Currently she coordinates the CiTUR Madeira research center and conducts research about carrying capacity in tourism, consumer behaviour in tourism, and digital marketing applied to tourism services. She is an author of scientific papers published in indexed international journals and of publications presented in peer-reviewed conferences. In 2020, she won a Literati Award for Excellence, supported by Emerald Publishing, with a highly commended paper. Also, she is the editor in chief of the *European Journal of Tourism, Hospitality and Recreation*, indexed in the Web of Science.

Contents

Preface

Post-COVID Tourism – Tendencies and Management Approaches is a critical theoretical-practical guidance for the tourism industry. Contextually, the monograph provides studies and reflections as a comprehensive roadmap, enabling readers to understand the challenges and transformative opportunities that have been defined after the COVID-19 pandemic.

Thus, the chapters focus on recovery, exploring innovative strategies and management approaches designed to foster resilience and sustainability. In this regard, considering the revitalization of travel destinations to the adoption of state-of-the-art technologies, each section provides an understanding of the multidisciplinary that constitute the future of tourism.

In sum, this book connects academia and practical application, offering tangible insights that somehow give the answer to the gap between theory and action.

Rui Alexandre Castanho and Mara Franco
CITUR Madeira - Centre for Tourism Research, Development and Innovation,
Madeira, Portugal

José Manuel Naranjo Gómez
University of Extremadura,
Badajoz, Spain

Chapter 1

Introductory Chapter: Post-COVID Tourism – Tendencies and Management Approaches – An Introduction

Rui Alexandre Castanho, José Manuel Naranjo Gómez and Mara Franco

1. Introduction

In the last few decades, the thematic literature and the evidence worldwide show that tourism activity is highly relevant and vital for numerous reasons. First, it has a significant economic impact. Tourism generates revenue through various channels such as accommodations, transportation, food and beverage services, entertainment, and shopping [1–3]. This tourist spending supports local businesses, creates job opportunities, and stimulates regional economic growth [4, 5].

Second, tourism facilitates cultural exchange. It allows people worldwide to experience and appreciate diverse cultures, traditions, and ways of life. Interactions between tourists and locals promote tolerance and respect for cultural diversity, fostering global understanding [3, 6].

Furthermore, tourism plays a crucial role in preserving heritage [3, 7]. Tourists often visit historical sites, natural landscapes, and cultural treasures, and the income generated from tourism can be used for their restoration and maintenance. This ensures the sustainability of these attractions for future generations to enjoy [8].

Tourism also raises environmental awareness [8, 9]. Responsible and sustainable tourism practices can educate visitors about environmental issues and encourage conservation efforts [3]. Experiencing fragile ecosystems and natural wonders firsthand often leads to a greater appreciation for the environment and a desire to protect it [3, 8, 10].

Additionally, tourism drives infrastructure development. To cater to the needs of tourists, destinations invest in improving transportation networks, accommodations, sanitation facilities, and recreational amenities. This benefits tourists and enhances the overall quality of life for local residents [11, 12].

Moreover, tourism has social benefits. Festivals, events, and tourism-related activities bring communities together, fostering social cohesion and providing platforms for cultural expression. Local residents take pride in showcasing their community and engaging with visitors, creating a sense of belonging and community spirit [13–15].

Last, tourism promotes education and knowledge exchange. Visitors learn about historical events, artistic expressions, scientific discoveries, and local traditions. Likewise, locals learn from the experiences and perspectives of tourists, creating opportunities for mutual education and understanding [16].

Nevertheless, it is vital to acknowledge the potential negative impacts of tourism, such as over-tourism, environmental degradation, cultural commodification, and socioeconomic inequalities. Contextually, it is pivotal to maximize the positive aspects of tourism while minimizing the harmful—i.e., sustainable tourism practices, responsible planning, and community involvement are crucial.

2. Tourism and the COVID-19 pandemic: in brief

The COVID-19 pandemic has had a profound and far-reaching impact on the tourism industry worldwide. Governments implemented travel restrictions and lockdown measures to contain the spread of the virus, leading to a significant decline in tourism activity. International travel came to a standstill, and domestic travel was severely restricted, resulting in a drop in tourist arrivals [17–19].

The economic consequences of the pandemic on tourism have been severe. Many businesses, particularly small and medium-sized enterprises, faced closures, layoffs, and financial instability. The job losses and revenue declines had a cascading effect on the entire tourism supply chain and the broader economy [18–20].

Recovering from the pandemic has posed significant challenges for the tourism industry. Despite lifting some restrictions, travelers' confidence and behavior have been impacted. Uncertainty surrounding the virus, changing travel regulations, and ongoing health concerns influence travel decisions.

The tourism industry has demonstrated resilience and adaptability in response to the crisis. Businesses have implemented stringent health and safety protocols to instill confidence in travelers. Virtual tours and online experiences have gained popularity as alternatives to physical travel, allowing people to explore destinations remotely [21].

The pandemic has also led to a shift in travel preferences. Travelers now prioritize health and safety measures, outdoor activities, and destinations with lower population density. Domestic and local travel has gained prominence due to travel restrictions and a desire to support local economies [22].

Sustainable tourism practices have gained increased importance during the pandemic. There is a growing recognition of the need to balance economic growth with environmental and social considerations. Destinations now prioritize sustainability and resilience in their tourism strategies, aiming for a more responsible and sustainable recovery [22, 23].

Collaboration and coordination among stakeholders have become essential during this crisis. Governments, tourism organizations, businesses, and local communities are working together to develop recovery plans, share best practices, and provide support to those affected by the pandemic [3].

It is important to note that the situation remains dynamic, with the impact of the pandemic on tourism evolving as vaccination efforts progress and travel restrictions are adjusted. The recovery of the tourism industry will depend on the global control of the virus, the success of vaccination campaigns, and the restoration of travelers' confidence in resuming their travel plans.

3. Tourism in the post-COVID: evidences and forecasting

If we focus on the evidence and forecasting for tourism in the post-COVID period, this is subject to ongoing developments and uncertainties. Regardless, specific trends and considerations can provide insights into the potential direction of the industry:

I. Domestic and regional focus: In the early stages of recovery, domestic and regional travel has been a primary driver of tourism activity. This trend will likely continue as travelers may prefer destinations closer to home due to ongoing concerns and restrictions on international travel [24].

II. Gradual recovery of international travel: While the recovery of international travel may take longer, it is expected to rebound as vaccination rates increase and travel restrictions ease gradually. However, the pace and extent of recovery will depend on factors such as the global control of the virus, government policies, and traveler confidence [25].

III. Health and safety measures: Health and safety protocols will remain a priority post-COVID era. Travelers will continue to expect rigorous sanitation practices, social distancing measures, and the use of technology for contactless services. Implementing these measures will be crucial to restoring and maintaining traveler confidence [26].

IV. Sustainable and responsible tourism: The pandemic has amplified the importance of sustainability and responsible tourism practices. Travelers are increasingly seeking environmentally friendly options, supporting local communities, and prioritizing destinations that prioritize sustainability. Post-COVID tourism will likely significantly emphasize sustainable practices and destination management [3].

V. Digital transformation: The pandemic has accelerated the digital transformation of the tourism industry. Online booking platforms, contactless services, and virtual experiences have become prevalent. Technology integration will continue to shape the post-COVID tourism landscape with innovations such as touchless payments, digital health passports, and enhanced communication platforms [27].

VI. Resilience and adaptability: The ability of tourism businesses to adapt and innovate has been critical during the pandemic. Those that have demonstrated resilience by diversifying their offerings, implementing flexible policies, and focusing on customer needs are more likely to navigate the post-COVID recovery successfully [7].

Also, we should consider that several factors, including the progression of vaccination efforts (particularly in less developed countries and emerging markets), the emergence of new variants, government policies, and traveler behavior, will influence the trajectory of tourism in the post-COVID period. Ongoing monitoring, data analysis, and expert insights are necessary to provide more accurate and up-to-date evidence and forecasting for tourism in the post-COVID era.

4. Closing thoughts

The COVID-19 pandemic has unprecedentedly impacted the tourism industry, leading to widespread disruptions and challenges. However, it has also presented an opportunity for the industry to reassess and transform itself for a more sustainable and resilient future.

As the world moves toward recovery, the tourism sector must prioritize health and safety measures, adapt to changing traveler preferences, and embrace sustainability. Collaboration among stakeholders, including governments, businesses, and local communities, will be essential in developing recovery strategies and ensuring a responsible and inclusive tourism industry.

While the future of tourism may still hold uncertainties, there is optimism for the industry's gradual rebound and resurgence. As vaccination efforts progress and travel restrictions ease, the desire to explore new destinations and reconnect with the world will drive the recovery.

Ultimately, the post-COVID period provides an opportunity to build a more sustainable, inclusive, and resilient tourism industry. By prioritizing safety, sustainability, and responsible practices, we can create a tourism sector that benefits economies, respects the environment, preserves cultural heritage, and enriches the lives of travelers and local communities alike.

Acknowledgements

This study is a result of the Research Group CITUR-Madeira-Centre for Tourism Research, Development and Innovation).

Author details

Rui Alexandre Castanho[1,2*], José Manuel Naranjo Gómez[1,3] and Mara Franco[1]

1 CITUR Madeira—Centre for Tourism Research, Development and Innovation, Funchal, Madeira, Portugal

2 WSB University, Dabrowa Górnicza, Poland

3 University of Extremadura, Badajoz, Spain

*Address all correspondence to: acastanho@wsb.edu.pl

References

[1] Stephen P. Potential economic contribution of regional tourism development in China: A comparative analysis. International Journal of Tourism Research. 2015;**17**(3):303-312. DOI: 10.1002/jtr.1990

[2] Labrianidis L, Ferrão J, Hertzina K, Kalantaridis C, Piasecki B, Smallbone D. The Future of Europe's Rural Periphery. Final Report. 5th Framework Programme of the European Community. 2003

[3] Castanho RA, Couto G, Pimentel. Principles of sustainable tourism and cultural management in rural and ultra-peripheral territories: Extracting guidelines for its application in the Azores Archipelago. Cultural Management: Science and Education (CMSE). 2020;**4**:9-24. DOI: 10.30819/cmse.4-1.01. ISSN: 2512-6962

[4] Vanegas M, Croes R. Growth, development and tourism in a small economy: Evidence from Aruba. International Journal of Tourism Research. 2003;**5**:315-330. DOI: 10.1002/jtr.441

[5] Castanho RA, Couto G, Naranjo Gómez JM, Pimentel P, Carvalho C, Sousa Á et al. Evolutionary dynamics in Azorean landscapes: The land-use changes in forests and semi-natural areas in the archipelago from 1990 to 2018. In: Rocha Á, et al, editors. Chapter No: 24 in the Book: Trends and Applications in Information Systems and Technologies. World CIST 2021, AISC 1367. Basel, Switzerland: Springer Nature Switzerland AG; 2021. pp. 244-252. DOI: 10.1007/978-3-030-72651-5_24

[6] Balaguer J, Jordá M. Tourism as a long-run economic growth factor: The Spanish case. Applied Economics. 2010;**34**(7):877-884. DOI: 10.1080/00036840110058923

[7] Castanho RA, Santos C, Couto G. Creative tourism in islands and regional sustainable development: What can we learn from the pilot projects implemented in the Azores territory? Land. 2023;**12**:498

[8] Ulucak R, Yücel A, Koçak E. The process of sustainability: From past to present. Environmental Kuznets Curve (EKC). 2019;**1**:37-53

[9] Loures L, Naranjo Gómez JM, Castanho RA, Loures A. Benefits and limitations of public involvement processes in landscape redevelopment projects—Learning from practice. In: Regional Intelligence Spatial Analysis and Anthropogenic Regional Challenges in the Digital Age. Basel, Switzerland: Ed. Springer; 2020. pp. 29-48. DOI: 10.1007/978-3-030-36479-3_3

[10] Cabezas J, Martín Gallardo J, Naranjo Gómez JM, Fernadez-Pozo L, Castanho RA. Serviços Ecossistêmicos e Geodiversidade em Paisagens do Mediterrânneo. In: Chapter 15 in the UNICAMP (Eds.) Book: Novos Rumos Do Direito Ambiental. Um Olhar Para a Geodiversidade. Coord. Luciana Cordeiro de Souza-Fernandes, Alexandra Aragão e Artur Sá. Campinas, Brazil: Ed. UNICAMP; 2021. pp. 349-366. ISBN13: 978-65-86253-87-0

[11] Castanho RA, Naranjo Gómez JM, Vulevic A, Couto G. The land-use change dynamics based on the CORINE data in the period 1990-2018 in the European archipelagos of the Macaronesia region: Azores, Canary Islands and Madeira. ISPRS International Journal of Geo-Information. 2021;**10**:342

[12] Mahony K, Zyl J. The impacts of tourism investment on rural communities: Three case studies in South Africa. Development Southern Africa. 2002;**19**(1):83-103. DOI: 10.1080/03768350220123891

[13] Yesavdar U, Belgibayav A, Mersakylova G. The role of developing direction of international tourism in Kazakhstan. Bulletin of the National Academy of Sciences of the Republic of Kazakhstan. 2016;**2**:180-185

[14] Williams A. Introduction. In: Williams A, editor. Southern Europe Transformed-Political and Economic Change in Greece, Italy, Portugal and Spain. New York, Ed. New York, USA: Harper and Row; 1984. pp. 1-32

[15] Couto G, Pimentel P, Ponte JC, Gonçalves P, Duarte D, Arruda A, et al. Azores: Investor's Guide to Sustainable Tourism. Ponta Delgada: FLAD-Fundação Luso Americana para o Desenvolvimento; 2018

[16] Couto G, Castanho RA, Pimentel P, Sousa Á, Graça Batista M. The pivotal factors for the success of rural tourism activities in Islands. In: Proceedings of the VII International Forum on Management; 23-24 January 2023; Funchal, Portugal. Funchal, Portugal: Ed. UMa; 2023. pp. 48-49. ISBN 978-989-8805-77-5

[17] Couto, G, Castanho, RA, Pimentel, P, Barreto Carvalho, C, Sousa, Á. How SARS-CoV-2 crisis could influence the tourism intentions of Azores archipelago residents? A study based on the assessment of the public perceptions. In: World Review of Science, Technology and Sustainable Development; Special issue: The Global Economy, the Tourism Sector and Sustainable Development. Vol. 19. Geneva, Switzerland: Inderscience Enterprises Limited; 2023. pp. 133-148. ISSN 17412242

[18] Abellera, K, Castanho, RA. Analysis of the sustainable tourism development in Indonesia. In: Nuevas Estrategias para un Turismo Sostenible. Cizur Menor, Spain: Thomson Reuters Aranzadi; 2022. pp. 371-380. ISBN 978-84-1124-804-4

[19] Castanho RA. A pandemic crisis shocking us all: The COVID-19. Special Issue of: Reflexiones Sobre el Coronavirus y Sus Impactos in Revista Científica Monfragüe Resiliente—Scientific Journal. 2020;**xx Special Issue**:233-238

[20] Baixinho A, Santos C, Couto G, Albergaria ISd, Silva LSd, Medeiros PD, et al. Creative tourism on islands: A review of the literature. Sustainability. 2020;**12**:10313

[21] Santos C, Couto G, Soares de Albergaria I, Sampaio da Silva L, Damião Medeiros P, Maria Neves Simas R, et al. Analyzing pilot projects of creative tourism in an ultra-peripheral region: Which guidelines can be extracted for sustainable regional development? Sustainability. 2022;**14**:12787

[22] Rebola F, Loures L, Ferreira P, Loures A. Inland or coastal: That's the question! Different impacts of COVID-19 on the tourism sector in Portugal. Sustainability. 2022;**14**:16248

[23] Walker T, Lee T, Li X. Sustainable development for small island tourism: Developing slow tourism in the Caribbean. Journal of Travel & Tourism Marketing. 2021;**38**:1-15

[24] Graça Batista M, Couto G, Castanho RA, Sousa Á, Pimentel P, Barreto Carvalho C. The rural and nature tourism development potential in islands. Sustainability. 2022;**14**:5289

[25] Behradfar A, Castanho RA, Couto G, Sousa Á, Pimentel P. Analyzing

COVID-19 post-pandemic recovery process in Azores archipelago. Konuralp Medical Journal. 2022;**14**:206-218

[26] Santos E, Crespo C, Moreira J, Castanho RA. Brand and competitiveness in health and wellness tourism. In: Katsoni V, editor. Tourism, Travel, and Hospitality in a Smart and Sustainable World. IACuDiT 2022. Springer Proceedings in Business and Economics. Cham: Springer; 2023. DOI: 10.1007/978-3-031-26829-8_44

[27] Pilving T, Kull T, Suškevics M, Viira AH. Creating shared collaborative tourism identity in a post-communist environment. Scandinavian Journal of Hospitality and Tourism. 2021;**21**(3):313-340

Chapter 2

Sustainable Healthcare Tourism in the Post COVID 19 Era

G. Somu and Harshavardhan Sai Sadineni

Abstract

Health tourism in India offers high-quality medical treatments to foreign tourists at affordable prices. The COVID-19 pandemic significantly impacted the industry, leading to a decline in the number of medical tourists. This chapter discusses the government's response to the crisis and the opportunities for sustainable healthcare tourism in the post-pandemic era. The Indian government issued guidelines for health tourism operators, emphasizing sustainability. These guidelines aim to control the virus's spread while ensuring the industry's long-term sustainability. The article explores the potential for sustainable healthcare tourism in India post-COVID and how the industry can attract more foreign tourists by leveraging changing trends. The role of the government in managing the health tourism sector is to highlight and focus on sustainable practices. Measures to ensure the safety of foreign tourists seeking medical treatment in India are discussed, including sustainable infrastructure development, responsible waste management, and resource conservation. By incorporating sustainable practices, the health tourism industry can reduce its environmental impact and contribute to the well-being of local communities.

Keywords: sustainable healthcare tourism, COVID-19, Indian healthcare, government guidelines, opportunities

1. Introduction

Health tourism, also known as medical tourism, involves setting foot in another country to receive medical treatment, often at a lower cost than in one's home country. In recent years, health tourism has become an important industry in India due to the country's world-class healthcare facilities, skilled doctors, and cost- effective treatments. Health tourism brings in revenue for the country and helps boost its economy.

The growth of health tourism in India has been significant over the years. The medical tourism industry worldwide was valued at around US$16.761 million and is expected to reach US$27.247.6 million by 2024, with a compound annual growth rate [CAGR] of 8.5% during the period 2019–2023. According to reports from FICCI and IMS Health India, India holds approximately 18% of the global medical tourism market share [1]. India's healthcare infrastructure has been steadily improving, and the country is now home to many world-class hospitals, clinics, and healthcare centers. In addition, the availability of skilled doctors and healthcare professionals in the country has also contributed to the growth of health tourism in India.

There are many reasons why India has become a popular destination for health tourism. Firstly, the cost of medical treatment in India is significantly lower than in countless other countries. Patients who travel to India for medical treatment can save money on their healthcare expenses [2]. Secondly, the quality of healthcare in India is excellent, with many hospitals and clinics equipped with state-of-the-art technology and facilities.

The government has introduced various policies and initiatives to make it easier for foreign patients to access medical treatment in India. For example, the government has simplified the visa process for medical tourists and introduced measures to ensure that healthcare providers in the country maintain high standards of quality and safety.

2. Impact of the pandemics on the healthcare tourism of India

In recent years, India has witnessed the impact of various pandemics on its healthcare tourism industry. The outbreaks of diseases like H1N1 influenza in 2009, Ebola virus during 2014–2016, and Zika virus in 2015–2016 have significantly affected the flow of medical tourists to the country, posing challenges to this crucial sector.

During these pandemics, healthcare tourism experienced a decline as potential travelers became increasingly concerned about the risk of contracting these diseases and the overall safety of medical facilities [3]. Governments worldwide issued travel advisories, urging their citizens to avoid non-essential travel to affected regions, including India, which reduced the number of international patients seeking medical treatment in the country.

The fear of infection and the uncertainty surrounding the pandemic resulted in many medical tourists canceling or postponing their scheduled treatments. The perceived risks associated with traveling for healthcare caused potential patients to rethink their decisions, opting to seek medical services locally rather than traveling abroad. This shift in patient behavior directly impacted the revenue generated from healthcare tourism in India.

Furthermore, during pandemics, healthcare providers and facilities were compelled to divert their attention and resources toward managing and containing the spread of diseases within their populations. The sudden influx of domestic patients suffering from pandemic-related illnesses strained the healthcare infrastructure, making it challenging to cater to medical tourists' needs effectively. This diversion of resources further hampered the delivery of healthcare services to international patients.

The impact of these pandemics on healthcare tourism served as a wake-up call, highlighting the sector's vulnerability to global health crises. It underscored the importance of implementing robust infection control measures and maintaining trust in the healthcare system. The ability to effectively communicate safety protocols and reassure potential medical tourists about the preparedness of healthcare facilities became critical.

As India gradually recovers from these pandemics, the healthcare tourism industry must rebuild trust and assure international patients of the safety and quality of healthcare services, achieved through a multi-faceted approach that encompasses enhanced safety protocols, stringent infection control measures, and transparent communication about the preparedness of healthcare facilities to handle any future health crises.

First and foremost, healthcare providers must prioritize implementing stringent infection control measures, including comprehensive training for healthcare personnel in handling contagious diseases and following strict hygiene protocols. Regular and thorough sanitization of medical facilities, proper waste management, and the availability of personal protective equipment [PPE] are crucial aspects of ensuring a safe healthcare environment.

Moreover, it is imperative to develop and enforce standardized guidelines and regulations specific to healthcare tourism, which address the challenges posed by pandemics. These guidelines should be regularly updated to align with international standards and cover areas such as patient screening, quarantine protocols, and emergency response mechanisms. By adopting and adhering to these guidelines, India can demonstrate its commitment to providing safe and reliable healthcare services to international patients.

Transparent and effective communication is pivotal in rebuilding trust and attracting medical tourists. Healthcare tourism stakeholders, including hospitals, clinics, and tourism boards, should actively engage in promotional activities highlighting safety measures, including showcasing the success stories of patients who received treatment in India during previous pandemics, emphasizing the high standards of care and the positive outcomes.

Public and private collaboration is essential to address the impact of pandemics on healthcare tourism effectively. Government agencies should work closely with healthcare providers and tourism authorities to develop strategies that mitigate risks and promote India as a safe and reliable destination for medical treatments, involving joint marketing campaigns, financial incentives for healthcare facilities, and the establishment of dedicated task forces to manage crises and support healthcare tourism.

3. Covid-19 impact on health tourism

Before the COVID-19 pandemic, the health tourism industry in India was booming. The country was a popular destination for medical tourists worldwide, particularly from the United States, the United Kingdom, and the Middle East [4]. India's healthcare system has a reputation for providing high-quality medical treatments and procedures at a fraction of the cost in developed countries.

Several types of medical treatments and procedures were popular among foreign tourists in India. These included cardiac surgery, orthopedic surgery, cosmetic surgery, and cancer treatment. Patients chose India for these treatments because of the country's skilled doctors and healthcare professionals, and affordable costs. In addition, India's diverse culture, rich history, and warm hospitality made it an attractive destination for medical tourism (**Figure 1**).

The COVID-19 pandemic has significantly impacted the health tourism industry in India. The government has implemented travel restrictions, quarantine measures, and lockdowns. These measures have made it difficult for medical tourists to travel to India for treatment. In addition, the pandemic has led to a shortage of medical supplies, including drugs and vaccines, and a strain on the healthcare infrastructure (**Figure 2**) [6].

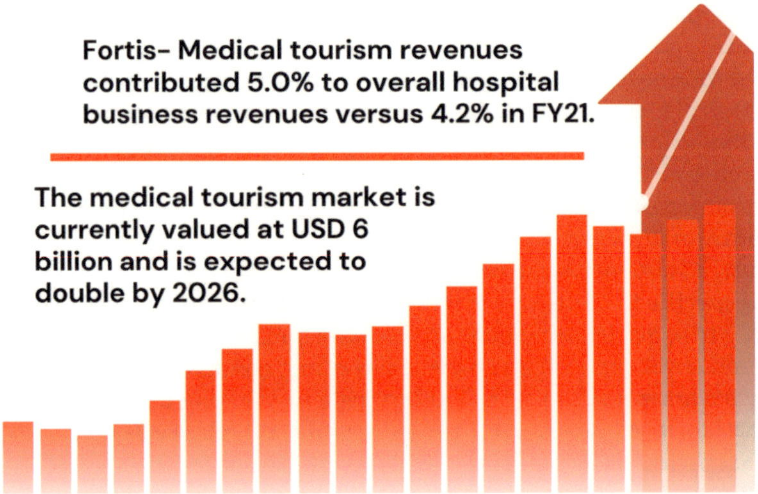

Figure 1.
Market value with examples [3, 5].

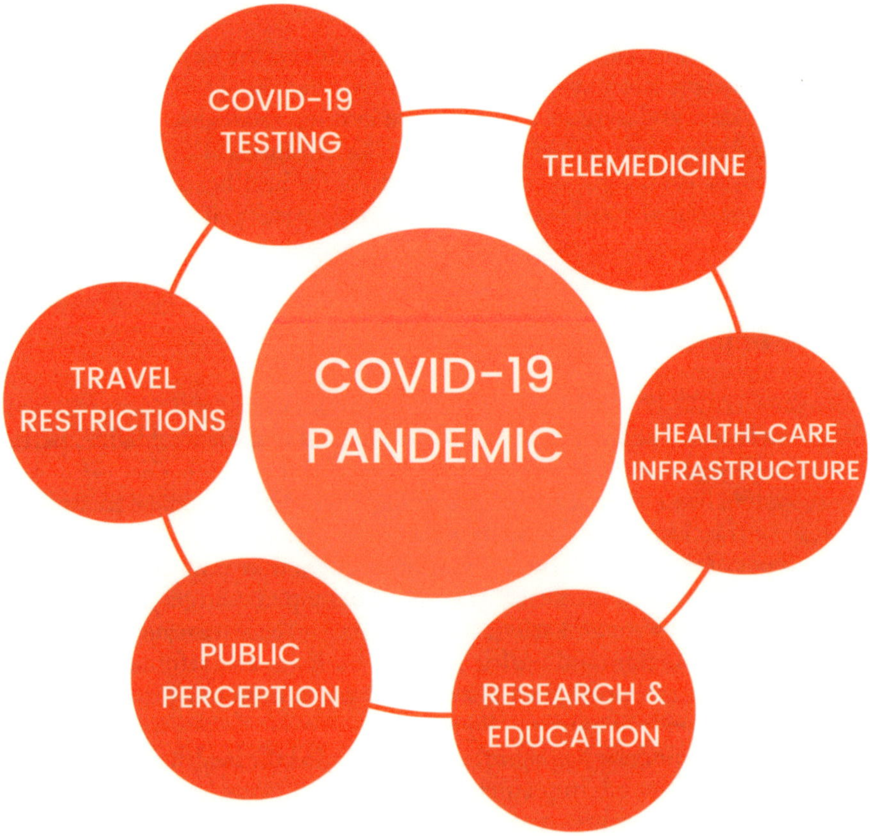

Figure 2.
Dimensions of medical tourism affected by the COVID-19 pandemic.

The government has taken various measures to control the spread of the virus and support the healthcare sector during the pandemic. These measures include increasing the production of medical supplies, providing financial support to healthcare workers, and launching a nationwide vaccination drive. The government has also introduced guidelines for health tourism operators to ensure the safety of domestic and foreign patients seeking medical treatment in the country.

The pandemic's impact on India's health tourism industry has been significant. The industry has experienced a sharp decline in revenue, with many hospitals and clinics reporting a drop in foreign patients. However, the industry has also adapted to the changing circumstances by offering telemedicine services and online consultations to patients who cannot travel to India for treatment.

4. Role of healthcare professionals in healthcare tourism during the pandemic

During the COVID-19 pandemic, healthcare professionals have played a vital role in reaching out to local and global patients, including those seeking healthcare services through medical tourism. Their expertise, dedication, and adaptability have been crucial in ensuring the safety and well-being of patients in these challenging times.

Locally, healthcare professionals have been at the forefront, having worked diligently to provide timely and accurate information about the virus, preventive measures, and treatment options to domestic and international patients. By staying updated with the latest research and guidelines, healthcare professionals have effectively addressed the concerns and queries of medical tourists.

In medical tourism, a crucial pillar was communicating with international patients and facilitating their healthcare needs. They have used various telemedicine technologies to provide virtual consultations, assess patient conditions, and offer medical advice [7]. This approach has allowed medical tourists to receive expert opinions and guidance from healthcare professionals without demanding physical travel, ensuring continuity of care.

Healthcare professionals have also been actively coordinating travel logistics and ensuring the seamless transfer of medical records and test results for international patients. By collaborating with healthcare tourism facilitators, they have helped streamline the process of accessing healthcare services in India, including assisting with visa arrangements, coordinating appointments, and providing the necessary documentation to facilitate medical treatments.

Moreover, stringent infection control protocols have been implemented to ensure the safety of international patients seeking medical care. They have followed international guidelines for hygiene practices, personal protective equipment [PPE] usage, and social distancing measures to create a safe healthcare environment.

These measures have protected patients and instilled confidence among medical tourists regarding the quality and safety of healthcare services in India (**Figure 3**).

Additionally, unique challenges of the medical tourists were recognized during the pandemic. They have provided personalized support and guidance to patients, addressing concerns related to travel restrictions, quarantine requirements, and changes in treatment plans. This level of care and attention has been crucial in maintaining a strong bond of trust between healthcare professionals and medical tourists.

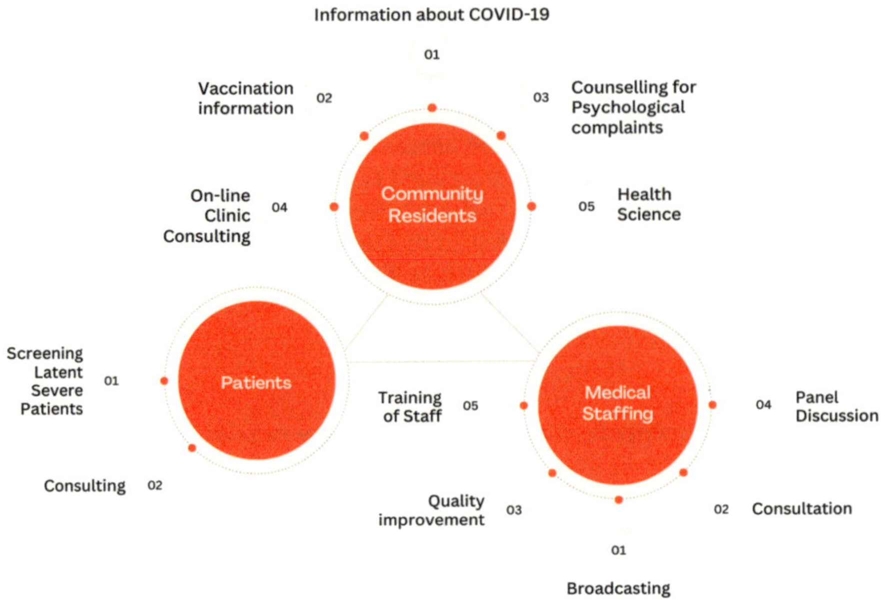

Figure 3.
Roles and benefits during the pandemic.

Furthermore, vaccination drives have been instrumental in ensuring the safety and well-being of local communities and international patients. By actively promoting vaccination and providing accurate information about vaccine efficacy and safety, healthcare professionals have contributed to the resumption of healthcare tourism by instilling confidence in potential medical tourists.

5. Opportunities and measures taken post-Covid

With the availability of vaccines and the easing of travel restrictions, the health tourism industry has bounced back post-COVID. One of the changing trends in the health tourism industry post-COVID is the increased demand for wellness tourism. Wellness tourism involves traveling to a destination to maintain or improve one's health and well-being. India, with its rich Ayurveda, yoga, and meditation history, is a natural destination for wellness tourism [8]. The Indian government can leverage this trend by promoting India as a destination for wellness tourism, highlighting the benefits of Ayurveda, yoga, and meditation, and supporting wellness tourism operators.

Another trend in the health tourism industry post-COVID is the increased use of technology. With the pandemic highlighting the importance of telemedicine and online consultations, health tourism operators can leverage technology to provide patients with remote consultations and follow-up care. The Indian government can support this trend by providing incentives and support to health tourism operators who adopt the technology.

The Indian government is crucial in managing the health tourism industry post-COVID. Various measures are in place for the safety of foreign tourists seeking medical treatment in India. These measures include issuing guidelines for health tourism operators, providing financial assistance to hospitals and clinics for upgrading their facilities, and launching a certification program for healthcare facilities to ensure compliance with international standards.

A Medical Value Travel Facilitation Portal in conjunction with the Heal in India and Heal by India Programs, is an online platform that provides information on healthcare facilities, medical treatments, and visa procedures to foreign tourists seeking medical treatment in India [9]. The government has also launched a marketing campaign to promote India as a destination for medical tourism, highlighting the benefits of the country's healthcare system and its affordable costs.

Several health tourism operators in India have followed government guidelines to ensure the safety of foreign tourists seeking medical treatment in India. One example is the Manipal Hospitals, which has implemented strict safety protocols, including mandatory COVID-19 testing, social distancing measures, and contact tracing. Manipal Hospitals has also adopted technology, providing patients with online consultations and follow-up care (**Figure 4**) [10].

Niti Aayog recognizes medical value travel as a key growth driver and significant forex earner [11]. Medical value travel benefits India by boosting the economy through foreign patient spending, enhancing healthcare reputation, promoting knowledge exchange, improving medical infrastructure, creating employment opportunities, and contributing to overall economic growth and global recognition.

6. Technology enabling during the pandemic for promoting health tourism

As countries recover from the pandemic and travel restrictions ease, technology has emerged as a crucial enabler for promoting health tourism in India in the post-pandemic era. With a focus on safety, convenience, and accessibility, technology

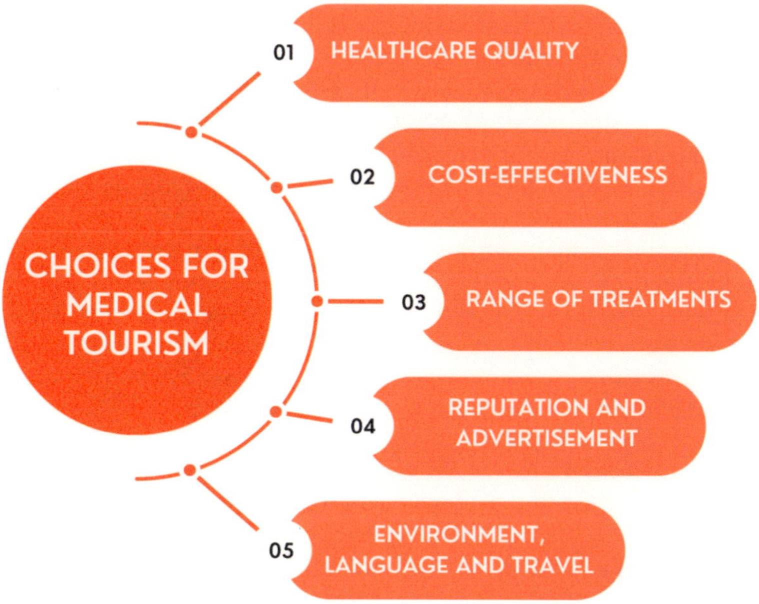

Figure 4.
Factors affecting the choice of destination for medical tourism.

solutions have transformed how medical travelers engage with healthcare providers and access healthcare services.

Telemedicine and virtual consultations have gained tremendous popularity during the pandemic. Through secure video conferencing platforms, patients can discuss their medical conditions, seek expert opinions, and even receive follow-up consultations from the comfort of their homes. This convenience not only saves time and costs for medical tourists but also allows them to receive timely medical advice and access specialized healthcare services in India.

Technology has facilitated the digitization of medical records, making it easier for medical tourists to access and share their medical history with healthcare providers in India. EMRs enable seamless information exchange between healthcare professionals, ensuring continuity of care and reducing the risk of errors. With secure electronic platforms, medical tourists can securely store and access their medical records, ensuring that healthcare providers have comprehensive and accurate information for effective diagnosis and treatment.

Dedicated health tourism portals and mobile applications have emerged as centralized platforms, providing medical tourists with comprehensive information and resources. These platforms offer various services, including information on healthcare providers, treatments, costs, and patient testimonials. Medical tourists can browse through different providers, compare services, read reviews, and make informed decisions about their healthcare journey in India. These platforms also facilitate online appointment booking, visa assistance, and accommodation arrangements, streamlining the entire process for medical tourists.

One such platform is HealthTrip, which is an innovative medical tourism platform that leverages advanced technologies like artificial intelligence and virtual reality to provide personalized treatment options. It offers comprehensive packages that include medical consultations, treatment planning, concierge services, and post-treatment care, ensuring a seamless and hassle-free experience for patients traveling abroad for medical purposes [12].

Healthcare providers and medical tourism agencies leverage social media platforms, search engine optimization [SEO], and targeted online advertisements to reach potential medical tourists globally. They highlight their expertise, state-of-the-art facilities, and success stories to build trust and attract patients. Additionally, online reputation management ensures that positive patient experiences are shared through reviews and testimonials, further promoting India as a preferred destination for health tourism.

With the help of data analytics, healthcare providers can gather valuable insights into patient preferences, treatment outcomes, and satisfaction levels [13]. By analyzing data, healthcare facilities can tailor their services to meet the specific needs of medical tourists. Personalized care, customized treatment plans, and individualized patient experiences enhance the overall satisfaction of medical tourists and encourage them to recommend India as a destination for health tourism.

7. Future plan

One of the challenges to health tourism in India is the perception of the country's healthcare system. While India has world-class healthcare facilities and skilled healthcare professionals, the perception of the country's healthcare system is often negative due to poor sanitation and hygiene, lack of transparency in pricing, and delays in obtaining medical visas. The quality of healthcare services must be improved, and must address the concerns of foreign tourists, and provide transparent pricing and visa procedures.

Another challenge is the need for more awareness among foreign tourists about the availability and quality of healthcare services. Many foreign tourists need to be made aware of the medical treatments and procedures available in India or the affordable costs of these treatments. The Indian government and health tourism operators must launch marketing campaigns to promote India as a destination for health tourism, highlighting the benefits of the country's healthcare system and providing information on medical treatments and procedures available in India.

The need for more infrastructure and facilities in some areas of India is another challenge. While major cities such as Delhi, Mumbai, and Bangalore have world-class healthcare facilities, many other regions need more infrastructure and facilities to attract foreign tourists. Financial and technical assistance must be provided to hospitals and clinics in these regions to upgrade their facilities and provide high- quality healthcare services.

The future of health tourism in India is bright, with several opportunities for growth in the coming years. One of the growth opportunitics is mcdical tourism, which involves traveling to a destination for medical treatments and procedures. The Indian government and health tourism operators can leverage this opportunity by promoting India as a destination for medical tourism, highlighting the benefits of the country's healthcare system, and providing high-quality healthcare services to foreign tourists. Another opportunity for growth in the health tourism industry in India is wellness tourism. India, with its rich Ayurveda, yoga, and meditation history, is a natural destination for wellness tourism.

The "Atithi Devo Bhava" principle forms the fundamental essence of Indian hospitality, encouraging hosts to treat guests with utmost respect akin to divinity. This ethos guides India's tourism and hospitality sector to welcome global tourists and warmly offer them an unforgettable experience. With its vast landscapes, diverse culture, rich heritage, and significant historic sites, India is a remarkable destination for travelers worldwide. Furthermore, India's reputation for producing exceptional doctors adds to its appeal as a medical tourism hotspot. This advantage makes India a preferred choice for individuals seeking medical treatments abroad.

Furthermore, four significant trends will shape the future of wellness tourism and India's position in the industry. These include Ayurveda, robotic surgeries, surgical camps, and startups. These trends contribute to the growth and development of India's medical tourism sector, offering unique and advanced healthcare solutions. Prime Minister Narendra Modi recently expressed his vision of India becoming a global medical tourism hub during the Global Investment in Ayush Summit. His passionate remarks highlighted the potential for India to save and serve millions of lives each year, earning it a prominent position in the evolving geopolitical landscape [14].

Moreover, India's healthcare system has gained recognition, with numerous politicians and government officials from Africa and Asia expressing gratitude for Indian healthcare providers who have rescued the lives of their loved ones, forming a valuable asset for India in the dynamic realm of international politics.

Additionally, insurance presents a significant opportunity for growth. Indian insurance companies should actively explore avenues to offer health insurance to foreigners, tapping into the potential of this market segment.

8. Conclusion

Health tourism, also known as medical tourism, has emerged as a significant industry in India, driven by the country's world-class healthcare facilities, skilled doctors, and cost-effective treatments. India's healthcare industry has experienced substantial growth over the years, with the country's healthcare infrastructure improving. The availability of affordable medical treatment in India has made it a popular destination for patients seeking high-quality healthcare at lower costs compared to their home countries.

However, the health tourism industry in India has faced challenges due to various pandemics, including the H1N1 influenza, Ebola virus, Zika virus, and, most recently, the COVID-19 pandemic. These outbreaks have resulted in a decline in the number of medical tourists traveling to India, as concerns about the risk of infection and the overall safety of medical facilities increased. Travel advisories and government restrictions have further impacted international patients flow.

The COVID-19 pandemic, in particular, has significantly impacted the health tourism industry in India. Travel restrictions, quarantine measures, and lockdowns have made it difficult for medical tourists to access healthcare services in the country.

The diversion of resources toward managing the pandemic within the domestic population has strained the healthcare infrastructure, making it challenging to cater to international patients' needs effectively. As a result, the industry has witnessed a decline in revenue and a shift in patient behavior, with many opting for local health-care services instead.

Healthcare professionals have played a crucial role in communicating with international patients and providing virtual consultations. Telemedicine and online platforms have become essential tools in delivering healthcare services remotely, ensuring continuity of care, and minimizing the risk of exposure to the virus.

As countries recover from the pandemic and travel restrictions ease, technology has emerged as a vital enabler for promoting health tourism in India during the post- pandemic era. Telemedicine and virtual consultations have gained popularity, offering convenience and timely access to specialized healthcare services for medical tourists. The digitization of medical records has facilitated seamless information exchange, ensuring continuity of care and accurate diagnosis. Dedicated health tourism portals and mobile applications have centralized resources and streamlined the healthcare journey for patients. Platforms like HealthTrip leverages advanced technologies to provide personalized treatment options and comprehensive packages for medical tourists.

To attract more international patients, healthcare providers and medical tourism agencies utilize social media, search engine optimization, and online advertisements to showcase their expertise and facilities. Online reputation management helps promote positive patient experiences and build trust. Data analytics plays a crucial role in gathering insights and tailoring services to meet the specific needs of medical tourists, enhancing overall satisfaction.

While the future of health tourism in India holds immense potential, there are challenges to overcome. Improving the perception of India's healthcare system, creating awareness among foreign tourists, and enhancing infrastructure and facilities in certain regions is crucial. Transparent pricing, efficient visa procedures, marketing campaigns, and infrastructure development will contribute to India's health tourism growth.

In this context, India's hospitality ethos, "Atithi Devo Bhava," which treats guests with the utmost respect, aligns with the country's efforts to welcome and serve international patients. With its diverse culture, rich heritage, and world-class healthcare facilities, India is a remarkable medical and wellness tourism destination. As expressed by Prime Minister Narendra Modi, the vision of India becoming a global medical tourism hub underscores the country's commitment to serving and saving lives, further solidifying its position in the evolving geopolitical landscape.

Looking ahead, the health tourism industry in India has a promising future. Medical tourism can leverage India's advanced healthcare services and attract international patients seeking affordable, high-quality treatments. Additionally, India's heritage in Ayurveda, yoga, and meditation positions it as a natural destination for wellness tourism. By leveraging technology, improving perceptions, raising awareness, and enhancing infrastructure, India can cement its position as a preferred destination for health tourism and contribute to the country's economic growth and global recognition.

Conflict of interest

The authors declare no conflict of interest.

Author details

G. Somu[1*] and Harshavardhan Sai Sadineni[2]

1 Department of Hospital Administration, Kasturba Medical College Manipal, Manipal Academy of Higher Education, Manipal, Karnataka, India

2 Kasturba Medical College Manipal, Manipal Academy of Higher Education, Manipal, Karnataka, India

*Address all correspondence to: somu.g@manipal.edu

References

[1] Kapadia D, Dongre P, Mahadevan R. Strategies for mitigating the effects of a pandemic on Indian medical tourism. Cardiometry. 2021;**20**:144-150

[2] JosephMarloweArticle.pdf [Internet]. [cited 2023 May 20]. Available from: http://www.medretreat.com/templates/UserFiles/Documents/JosephMarloweArticle.pdf

[3] Sultana S, Haque A, Momen A, Yasmin F. Factors affecting the attractiveness of medical tourism destination: An empirical study on India-review article. Iranian Journal of Public Health. 2014;**43**(7):867-876

[4] VISION 7.pdf [Internet]. [cited 2023 May 20]. Available from: https://dgcmukerian1.org/popup/vision/VISION%207.pdf#page=9

[5] FHL+-+Q4FY22+%26+FY22+Press+Release.pdf [Internet]. [cited 2023 May 20]. Available from: https://www.fortishealthcare.com/drupal-data/investors/FHL%2B-%2BQ4FY22%2B%2526%2BFY22%2BPress%2BRelease.pdf

[6] Impact of the Coronavirus Pandemic on the Supply Chain in Healthcare | British Journal of Healthcare Management [Internet]. [cited 2023 May 20]. Available from: https://www.magonlinelibrary.com/doi/abs/10.12968/bjhc.2020.0047

[7] Iyengar K, Upadhyaya GK, Vaishya R, Jain V. COVID-19 and applications of smartphone technology in the current pandemic. Diabetes and Metabolic Syndrome: Clinical Research & Reviews. 2020;**14**(5):733-737

[8] Smith M, Puczko L. Health and Wellness Tourism. London: Routledge; 2008. p. 399

[9] www.ETGovernment.com. "One Step" Portal to be Launched to Provide Credible Info on Medical Facilities and Treatment in India: Mansukh Mandaviya - ET Government [Internet]. ETGovernment.com. [cited 2023 May 20]. Available from: https://government.economictimes.indiatimes.com/news/governance/one-step-portal-to-be-launched-to-provide-credible-info-on-medical-facilities-and-treatment-in-india-mansukh-mandaviya/91536510

[10] www.ETCFO.com. Teleconsulting Boom to Continue Post Pandemic: Manipal Hospitals CFO - ETCFO [Internet]. ETCFO.com. [cited 2023 May 20]. Available from: https://cfo.economictimes.indiatimes.com/news/teleconsulting-boom-to-continue-post-pandemic-manipal-hospitals-cfo/80873060

[11] Medical Tourism Forms a Promising Component of Apollo Hospitals Portfolio: Sangita Reddy [Internet]. Business Insider. [cited 2023 May 20]. Available from: https://www.businessinsider.in/business/news/medical-tourism-forms-a-promising-component-of-apollo-hospitals-portfolio-sangita-reddy/articleshow/73055701.cms

[12] About HealthTrip - Leading Medical Travel Services in India [Internet]. [cited 2023 May 20]. Available from: https://healthtrip.com/about-us

[13] Batko K, Ślęzak A. The use of big data analytics in healthcare. Journal of Big Data. 2022;**9**(1):3

[14] Ahmed D. India to Emerge as Largest Destination for Medical Tourism. The Times of India [Internet]. [cited 2023 May 21]; Available from: https://timesofindia.indiatimes.com/blogs/voices/medical-tourism-destination-india/

Chapter 3

Tourism and Sustainable Landscapes: A Portuguese Case

Rossana Santos and Rui Alexandre Castanho

Abstract

Nowadays, the need for sustainable development and growth is considered as a maximum. Therefore, in Mediterranean territories, this issue is not dissociable as well. Nevertheless, this concept of sustainability, when applied to the environmental sphere, is not always fully considered and understood—as is the case of the landscape's sustainability. Contextually, the present chapter intends to provide evidence of what way migrants and tourism businesses can change Mediterranean territories into sustainable landscapes. A Portuguese case study of 5,157 inquired migrants revealed that a significant percentage of them wish to return to their home country of origin to manage a tourism business and practice agriculture for self-consumption. This is owing to their tradition of practicing agriculture, where they generally do not employ insecticides and use simple techniques, as well as to the rural organization of the house and other cultivated lands they have scattered in their place of origin. Their similar experiences and migratory characteristics also lead us to argue there is a strong probability that migrants from Southern Europe and Mediterranean countries can also contribute to tourism development and sustainable landscapes in those countries.

Keywords: tourism, territorial management, sustainable landscapes, Portuguese emigrants, Southern Europe and Mediterranean countries

1. Introduction

This material is devoted to the analysis of the impact of Portuguese return migration on the development of tourism and sustainable landscapes. As such, it starts contextualizing the potential of return migration movement, entrepreneurship and tourism on sustainable development and sustainable landscapes, especially in the countries of Southern Europe and the Mediterranean.

On this point, in the following sections, the results obtained in a study developed in Portugal through the application of a questionnaire survey of 5,157 Portuguese emigrants will be illustrated. It is shown that the emigrants with active age and a house in a rural area would like to return, have sufficient capital to invest, have a job in tourism in Portugal and would also like to reconcile tourism with traditional agriculture, mainly for self-consumption. In the final section, this chapter will also indicate some directions for future studies in countries of Southern Europe and the Mediterranean with similar migratory characteristics and historical profiles.

2. Return migration, tourism and sustainable development and landscapes

European rural territories face various socio-economic challenges, including unemployment, lack of income opportunities, negative migratory balances, emigration and aging population. The countries whose majority of its regions are less competitive are those where agriculture plays a relevant role [1]. On this point, Labrianidis and Thanassis [1] argue that entrepreneurship, associated with small- and medium-sized businesses, is one way to achieve rural development in low-density territories of Europe once it allows increasing employment and income. Also, the Organization for Economic Cooperation and Development [2] has argued that emigrants can return with training, professional experience and financial and social capital acquired abroad.

Thus, the emigration-return movement, except in its initial phase (the departure), has several potentialities in terms of rural development, since emigrants preferably head to the regions of departure, which are the most deprived [3–16]. However, in the countries of Southern Europe and the Mediterranean, there has been no impact of the return of emigrants on the development of their places of origin. They have returned with conservative ideas, reflected in the acquisition of a small parcel of land and a new house, as well as business objectives based only on creating small-scale companies linked to the services sector, particularly shops and bars [17, 18].

In Portugal, the excessive fragmentation of property is an important limitation to national agriculture development [19]. In this way, the link between agriculture and tourism in rural areas presents itself not only as an opportunity to diversify the economy but also as a strategy for raising income for rural households [20, 21]. Scientific literature has shown that tourism can be a viable development strategy in rural territories due to its multiplier effect in creating employment and income in this sector and others of the economy that support or attract visitors and tourism companies [21–51]. Therefore, in non-competitive farms, tourism could become the main activity where agriculture, primarily for self-consumption, could function as a complementary activity, providing a substantial income supplement.

Nevertheless, prior to the COVID-19 outbreak, statistical data on international tourism for the previous three years, published by the World Tourism Organization, has shown that Europe—especially the sub-regions of Southern/Mediterranean and Western Europe—were the ones that evidenced the most pronounced seasonality of tourism demand in July, August and September. This scenario has been greatly encouraged by the demand for sun and sea tourism, concentrated in only three months of the year, along the coastline of the most appealing destinations. Being heavily dependent on few touristic markets, this behavior has been causing pressure on natural and cultural landscapes. On this point, and in the context of the COVID-19 pandemic, the UNWTO [52] has argued that rural tourism development is relevant because of the main motivations of these visitors for uncongested places, preferably with open-air experiences and activities. Thus, rural tourism development, return migration and entrepreneurship associated with small and medium-sized businesses present sustainable development and landscape opportunities for low-density territories in Southern Europe and Mediterranean countries.

3. Designing sustainable landscapes: in general terms

The thematic literature and the practical examples tell us, in general terms, that designing sustainable landscapes involves creating environments that balance human needs with ecological considerations, ensuring long-term resilience, functionality and esthetic appeal [53–56]. Contextually, we can name some fundamental principles and considerations for designing sustainable landscapes:

a. Assess the site: Begin by understanding the site's natural features, such as topography, soil composition, hydrology and existing vegetation. Evaluate any constraints or opportunities for sustainable design, such as sunlight exposure, water availability and wind patterns [57];

b. Preserve and restore: Prioritize the preservation and restoration of existing natural features, including native plants, wildlife habitats and water bodies. Protecting biodiversity and ecological processes is essential for the long-term health and resilience of the landscape [58, 59];

c. Water management: Implement effective water management strategies, such as rainwater harvesting, storm water management and water-efficient irrigation systems. Use permeable surfaces and green infrastructure to reduce runoff and promote groundwater recharge [57, 60];

d. Native plant selection: Opt for native plants adapted to the local climate and soil conditions. Native plants require less water, fertilizer and maintenance once established while also supporting local ecosystems and wildlife. Incorporate a diverse range of plants to enhance biodiversity and ecosystem services [58, 61];

e. Sustainable materials: Choose sustainable materials for hardscape elements such as pathways, walls and structures. Use recycled or locally sourced materials with low embodied energy. Consider the life cycle impacts, durability and maintenance requirements of materials to minimize environmental footprints [60];

f. Energy efficiency: Integrate energy-efficient design principles into the landscape. Use vegetation strategically for shading and wind protection to reduce energy consumption for heating and cooling. Incorporate renewable energy technologies, such as solar panels or small wind turbines, where feasible [62];

g. Ecological connectivity: Create linkages and corridors between different habitats to promote environmental connectivity and wildlife movement. Design green spaces that allow for the natural flow of plants, animals and ecological processes across the landscape [63];

h. Community engagement: Involve the local community in the design and maintenance of sustainable landscapes. Encourage educational opportunities and community involvement in stewardship activities. Consider cultural and social aspects to ensure that the landscape meets the needs and aspirations of the people who will use and benefit from it [64];

i. Maintenance and management: Develop a long-term management plan for the landscape to ensure its ongoing sustainability. Include measures for invasive species control, proper waste management, regular maintenance and monitoring of ecological health [57];

j. Adaptability and resilience: Design landscapes with adaptability in mind, considering climate change and potential future challenges. Incorporating resilient plant species, designing for water efficiency and creating spaces that can accommodate changing conditions [63] are just a few examples.

Besides, we should consider that designing sustainable landscapes is a dynamic and iterative process that requires collaboration, ongoing evaluation and adaptation. By considering ecological principles, resource efficiency and community engagement, we can create landscapes that are not only visually appealing but also contribute to the well-being of both humans and the natural environment [65].

4. Methodology

Bearing in mind the positive impact of return migration movement, entrepreneurship and tourism development on low-density territories, a study in Portugal, funded by the Foundation for Science and Technology (FCT), was developed in order to assess the contribution of the return of Portuguese emigrants on tourism development in Portugal [66]. The population of Portuguese emigrants is unknown, and the data were obtained through the application of a questionnaire survey to Portuguese emigrants totalling 5,157 filled and valid questionnaires. For the design of the questionnaire, nominal scales, checklists and ratio scales were used. Data analysis was carried out by means of quantitative descriptive analysis (frequencies and crossings) and correlation tests of variables, as well as a qualitative analysis through the application of Student's t-tests and ANOVA (analysis of variance) tests (Kruskal-Wallis and Mann-Whitney).

5. Results and discussion

The conclusions of the study described in the previous section carried out with 5,157 Portuguese emigrants scattered around the world [53] have shown that those aged 29–39 who have a house in a rural municipality (or less-developed area) in Portugal are the ones who would like to return and work in the tourism sector, as also have sufficient capital to invest in this sector in their place of origin (**Tables 1–3**).

The same study [53] also points out that emigrants also would like to have control over their own working hours in Portugal in order to reconcile tourism with traditional agriculture, mainly for self-consumption, since their house in the place of origin has vegetable gardens and/or other farmland scattered throughout the village, where they generally use simple techniques and do not employ insecticides (**Tables 4** and 5). Thus, these Portuguese emigrants can contribute to rural tourism

County of the house	Age			
	Less than 18 years old	Between 18 and 28 years old	Between 29 and 39 years old	Total
Highest centrality index	0.6%	2.1%	6.7%	9.3%
Lowest centrality index	0.3%	7.2%	15.9%	23.3%
Does not know/does not answer	1.3%	30.0%	36.1%	67.4%
Total	2.1%	39.2%	58.7%	100.0%

Source: Own elaboration.

Table 1.
Age of Portuguese emigrants vs. County of the house in Portugal.

County of the house	Most important factors for returning to Portugal								
	A	B	C	D	E	F	G	H	I
Highest centrality index	18.2%	19.4%	16.3%	18.6%		17.1%	10.5%	15.5%	16.0%
Smallest centrality index	81.0%	79.9%	83.3%	80.9%	81.3%	82.3%	89.3%	83.8%	84.0%
Does not know/does not answer	0.8%	0.7%	0.4%	0.5%	0.7%	0.6%	0.3%	0.7%	—

(A) Have a job opportunity; (B) Have income opportunities; (C) Have a rural lifestyle; (D) Live near my family in Portugal; (E) Existing infrastructures for living and investing in a business at place of origin; (F) Existing infrastructures for living and investing in a business at place of origin; (G) Have a house in place of origin; (H) Be able to exercise a paid activity, by my own, in the tourism sector; (I) None. Source: Own elaboration.

Table 2.
County of the house in Portugal vs. Most important factors for returning to Portugal.

County of the house in Portugal	Capital to invest in a business in Portugal				Total
	Yes	Maybe	No	Does not know/does not answer	
Highest centrality index	4.2%	7.2%	5.4%	4.7%	21.5%
Smallest centrality index	12.9%	24.4%	28.7%	11.3%	77.4%
Does not know/does not answer	0.1%	0.3%	0.2%	0.5%	1.1%
Total	17.2%	31.9%	34.3%	16.6%	100.0%

Source: Own elaboration.

Table 3.
County of the house in Portugal vs. Capital to invest in a business in Portugal.

%		Age of Portuguese emigrants		
		Less than 18 years old	Between 18 and 28 years old	Between 29 and 39 years old
If emigrants had a job in the tourism sector, in Portugal, they would like to control their own working hours due to several factors.	(a)	—	39.6%	**60.4%**
	(b)	—	33.5%	**66.5%**
	(c)	—	34.3%	**65.7%**
	(d)	—	**53.4%**	46.6%
	(e)	—	41.9%	58.1%
	(f)	—	41.3%	58.7%
	(g)	—	**45.3%**	54.7%
	(h)	—	41.3%	58.7%
	(i)	—	**50.0%**	50.0%
	(j)	—	41.0%	59.0%
	(l)	—	44.6%	55.4%
	(m)	—	42.1%	57.9%
	(n)	—	41.6%	58.4%
	(o)	**9.1%**	34.7%	56.2%

(A) I would also want to practice agriculture, primarily for self-consumption; (B) I would have a job in the area of tourism; (C) I want to have a rural lifestyle; (D) I have a hard time getting a permanent job; (E) I could have had more than one job; (F) I could better reconcile work with my family/social life; (G) It would be easier to comply with health commitments (among others); (H) I could devote more time to domestic work; (I) I would be less tired by the end of the day; J) I would have a higher level of job satisfaction; (L) I would be more productive at work; (M) I would have less intention of changing job; (N) I would have less pressure; (O) Did not know/did not answer. Source: Own elaboration.

Table 4.
Age vs. Portuguese emigrants would like to control their own working hours in Portugal due to a number of factors.

%		Age of Portuguese emigrants		
		Less than 18 years old	Between 18 and 28 years old	Between 29 and 39 years old
Portuguese emigrants could practice agriculture for self-consumption in Portugal, due to a number of factors	(a)	—	44.9%	55.1%
	(b)	—	39.3%	**60.7%**
	(c)	—	43.6%	56.4%
	(d)	—	37.5%	**62.5%**
	(e)	—	43.3%	56.7%
	(f)	4.8%	35.1%	60.2%

(A) The house in the place where I was born has vegetable gardens and groves; (B) I own other farmlands dispersed in the place where I was born; (C) When I am in Portugal I practice agriculture (especially for self-consumption); (D) I usually practice agriculture with simple techniques and not employ insecticides; (E) Agriculture for self-consumption allow me to increase disposable income; (F) Did not know/did not answer. Source: Own elaboration.

Table 5.
Age vs. Portuguese emigrants could practice agriculture for self-consumption in Portugal, due to a number of factors.

development and, at the same time, to sustainable landscapes in their place of origin, since the maintenance of small and very small farms, with little environmental impact, is a very important element for rural tourism destinations and sustainable landscapes. Furthermore, their houses in the place of origin are already built.

6. Final thoughts

This study has argued that the return migration, entrepreneurship associated with small- and medium-sized businesses, and tourism present opportunities for tourism development and sustainable landscapes in Portugal. This is corroborated by Portuguese emigrants' intentions of return, tourism investment and work, as well as their traditional agricultural taste and skills they put into practice in the house and small farms they have in the place of origin. Their similar experiences and migratory characteristics also lead us to argue there is a strong probability that emigrants from Southern Europe and Mediterranean countries can also contribute to tourism development and sustainable landscapes in those countries.

Thereby, return migration can positively affect sustainable development by facilitating knowledge transfer and enhancing local capacities. For example, returning migrants who have gained expertise in environmentally friendly practices or sustainable technologies can help promote sustainable development in their home regions. Additionally, remittances sent by return migrants can provide financial support for local entrepreneurship and investment in sustainable projects.

Also, tourism can contribute to sustainable development by generating income, creating job opportunities and fostering local entrepreneurship. It can also raise awareness about environmental conservation and cultural preservation. However, uncontrolled or poorly managed tourism can lead to environmental degradation, loss of cultural heritage and socio-economic imbalances.

In fact, landscapes are crucial in sustainable development as they provide the natural resources, ecosystems and cultural heritage that support human well-being. Sustainable development requires landscape conservation and wise management to ensure long-term viability. This includes protecting biodiversity, managing land and water resources sustainably, preserving cultural and historical sites, and promoting sustainable practices in various sectors such as agriculture, energy and tourism.

Sustainable development and landscapes are intertwined with return migration and tourism. Return migration can bring back knowledge and resources, contributing to sustainable landscape management. When practiced sustainably, tourism can help preserve landscapes by generating income for conservation efforts and raising awareness about their value. At the same time, sustainable landscapes and well-managed natural and cultural resources can attract tourists interested in responsible travel.

Considering the pronounced seasonality of tourism demand for sun and sea in the most appealing destinations of Southern Europe and Mediterranean countries, Portugal may constitute a pilot study in the sense that possible conclusions can be drawn by extrapolation to the same countries with similar migratory characteristics and historical profiles.

Acknowledgements

This study is a result of the Research Group CITUR–Madeira–Centre for Tourism Research, Development and Innovation).

Author details

Rossana Santos* and Rui Alexandre Castanho
CITUR Madeira—Centre for Tourism Research, Development and Innovation, Portugal

*Address all correspondence to: rossana.santos@staff.uma.pt

References

[1] Labrianidis L, Ferrão J, Hertzina K, Kalantaridis C, Piasecki B, Smallbone D. The Future of Europe's Rural Periphery. Final Report. Brussels: 5th Framework Programme of the European Community; 2003

[2] Organisation for Economic Cooperation and Development. Return migration: A new perspective. In: OCDE, editor. International Migration Outlook 2008. France: OECD; 2008. pp. 161-222

[3] Rocha-Trindade M. Comunidades emigrantes em situação dipolar: análise de três casos de emigração especializada para os E.U.A., para o Brasil e para França. Análise Social. 1976;**XII**:983-997

[4] Silva M, Amaro R, Clausse G, Conim C, Matos M, Pisco M, et al. Regresso, Emigração e Desenvolvimento Regional em Portugal. Lisboa: Instituto de Estudos para o Desenvolvimento; 1984

[5] Serrão J. Notas sobre emigração e mudança social no Portugal contemporâneo. Análise Social. 1985;**XXI**:995-1004

[6] Rocha-Trindade M. Fenómeno da Emigração em Portugal. Lisboa: Ed. Secretaria de Estado da Ciência e Tecnologia; 1992

[7] Ferrão J. Três Décadas de Consolidação do Portugal Demográfico. In: Barreto A, Preto C, editors. A Situação Social em Portugal, 1960-1995. Lisboa: Ed. Instituto de Ciências Sociais da Universidade de Lisboa; 1996. pp. 165-190

[8] Portela J, Nobre S. Entre Pinela e Paris: Emigração e regressos. Análise Social. 2001;**XXXVI**:1105-1146

[9] Wang W, Fan C. Success or failure: Selectivity and reasons of return migration in Sichuan and Anhui, China. Environment and Planning A. 2006;**38**(5):939-958. DOI: 10.1068/a37428

[10] Gonçalves M. Desenvolvimento em Meio Rural. Contributos da emigração e do regresso. Aplicação ao concelho de Boticas na região Barrosã [thesis]. Open University; 2007. Available from: http://hdl.handle.net/10400.2/778

[11] Démurger S, Xu H. Return migrants: The rise of new entrepreneurs in rural China. World Development. 2011;**39**(10):1847-1861

[12] Marchetta F. Return migration and the survival of entrepreneurial activities in Egypt. World Development. 2012;**40**(10):1999-2013. DOI: 10.1016/j.worlddev.2012.05.009

[13] Wahba J, Zenou Y. Out of sight, out of mind: Migration, entrepreneurship and social capital. Regional Science and Urban Economics. 2012;**42**(5):890-903. DOI: 10.1016/j.regsciurbeco.2012.04.007

[14] King R, Oruc N. Editorial introduction: Migration in the Western Balkans–trends and challenges. Journal of Balkan and Near Eastern Studies. 2019;**21**(1):1-10. DOI: 10.1080/19448953.2018.1532682

[15] Thanh N, Lebailly P, Dien N. Internal return migration in rural of Vietnam. Mediterranean Journal of Social Sciences. 2019;**10**(1):27-38

[16] Lewis J, Williams A. The economic impact of return migration in Central Portugal. In: King R, editor. Return Migration and Regional Economic

Problems. London: Ed. Croom Helm; 1986. pp. 100-128

[17] King R, Strachan A, Mortimer J, Gastarbeiter J. Go home: Return migration and economic change in the Italian mezzogiorno. In: King R, editor. Return Migration and Regional Economic Problems. London: Ed. Croom Helm; 1986. pp. 38-68

[18] Williams A. Introduction. In: Williams A, editor. Southern Europe Transformed-Political and Economic Change in Greece, Italy, Portugal and Spain. New York: Ed. Harper and Row; 1984. pp. 1-32

[19] Avillez F, Monke E. Small firms in northern and central Portugal. In: Monke E, Avillez F, Pearson S, editors. Small Farm Agriculture in Southern Europe. CAP Reform and Structural Change. London: Ashgate Publishing Ltd; 1998. pp. 1-6

[20] Cordovil F, Dimas B, Alves R, Baptista D. A Política Agrícola e Rural Comum e a União Europeia. 1st ed. Centro de Informação Europeia Jaques Delors. 2003

[21] Hughes D, Shields M. Revisiting tourism regional economic impact: Accounting for secondary household employment. Review of Regional Studies. 2007;**37**:186-206. DOI: 10.52324/001c.8294

[22] Bergstrom J, Cordell K, Ashley G, Watson A. Economic impact of recreational spending on rural areas: A case study. Economic Development Quarterly. 1990;**4**(1):29-39. DOI: 10.1177/089124249000400104

[23] Fleischer A, Felsenstein D. Support for rural tourism–does it make a difference? Annals of Tourism Research. 2000;**27**:1007-1024. DOI: 10.1016/S0160-7383(99)00126-7

[24] Balaguer J, Jordá M. Tourism as a long-run economic growth factor: The Spanish case. Applied Economics. 2010;**34**(7):877-884. DOI: 10.1080/00036840110058923

[25] Mahony K, Zyl J. The impacts of tourism investment on rural communities: Three case studies in South Africa. Development Southern Africa. 2002;**19**(1):83-103. DOI: 10.1080/03768350220123891

[26] Vanegas M, Croes R. Growth, development and tourism in a small economy: Evidence from Aruba. International Journal of Tourism Research. 2003;**5**:315-330. DOI: 10.1002/jtr.441

[27] Mazumder H, Ahmed M, Al-Amin Q. Estimating total contribution of tourism to Malaysia economy. International Journal of Business, Management and Social Sciences. 2009;**2**(3):146-159. Available from: http://www.ijbmss-ng.com/vol2-no3-2011.htm

[28] Chen K, Yang H. Appraising the economic impact of the opening up to mainland Chinese tourist arrivals' policy on Taiwan with a tourism—CGE model. Asia Pacific Journal of Tourism Research. 2010;**15**(2):155-175. DOI: 10.1080/10941661003629961

[29] Haddad E, Porsse A, Rabahy W. Domestic tourism and regional inequality in Brazil. Tourism Economics. 2013;**19**(1):173-186. DOI: 10.1080/13683500.2022.2126965

[30] Klytchnikova I, Dorosh P. Tourism sector in Panama: Regional economic impacts and the potential to benefit the poor. Natural Resources Forum. 2014;**32**(2):70-79. DOI: 10.1111/1477-8947.12019

[31] Lee C, Lee M, Yoon SH. Estimating the economic impact of convention and

exhibition businesses, using a regional input-output model: A case study of the Daejeon Convention Center in South Korea. Asia Pacific Journal of Tourism Research. 2013;**18**(4):330-353. DOI: 10.1080/10941665.2012.658414

[32] Snyman S. Household spending patterns and flow of ecotourism income into communities around Liwonde National Park, Malawi. Development Sothern Africa. 2013;**30**(4-5):640-658. DOI: 10.1080/0376835X.2013.832149

[33] Fodranov I, Kubickova V, Michalkova A. Measuring societal value of tourism: A new approach. Tourism. 2015;**63**(4):423-434. Available from: https://hrcak.srce.hr/149994

[34] Hyojin K, Byung-Gook K. Economic impacts of the hotel industry: An input-output analysis. Tourism Review. 2015;**70**(2):132-149. DOI: 10.1108/TR-11-2014-0056

[35] Stephen P. Potential economic contribution of regional tourism development in China: A comparative analysis. International Journal of Tourism Research. 2015;**17**(3):303-312. DOI: 10.1002/jtr.1990

[36] Crompton J, Ji Y, Dudensing R. Sources of variation in economic impact multipliers. Journal of Travel Research. 2016;**55**(8):1051-1064. DOI: 10.1177/00472875156172

[37] Yesavdar U, Belgibayav A, Mersakylova G. The role of developing direction of international tourism in Kazakhstan. Bulletin of the National Academy of Sciences of the Republic of Kazakhstan. 2016;**2**:180-185

[38] Croes R, Rivera M. Tourism's potential to benefit the poor: A social accounting matrix model applied to Ecuador. Tourism Economics. 2017;**23**(1):29-48. DOI: 10.5367/te.2015.0495

[39] Kim S, Miller C. An economic model comparison of EMSI and IMPLAN: Case of mistletoe marketplace. Tourism Economics. 2017;**23**(5):1124-1130. DOI: 10.1177/1354816616656420

[40] Ruzic P, Demonja D. Transformations in business & economics. Economic Impacts of Rural Tourism in Rural Areas of Istria. 2017;**16**(3):31-40. Available from: https://www.bib.irb.hr/913141

[41] Sánchez V, Fernandez M, Lara J. Economic impact of a religious and tourist event: A holy week celebration. Tourism Economics. 2017;**23**(6):1255-1274. DOI: 10.1177/135481661667

[42] Veretekhina S, Shinkareva O, Kozhaev J, Telepchenkova N, Kuznetsova E, Plekhanov N. Evaluation methodology of the multiplier effect for the region as the result of the cluster formation. Eurasian Journal of Analytical Chemistry. 2017;**12**:533-547. DOI: 10.12973/ejac.2017.00188a

[43] Yashalova N, Akimova M, Aleksandrovich R. Propects for regional development of industrial tourism in view of the analysis of the main economic indicators of Russian tourism industry. Economic and Social Changes-Facts Trends. 2017;**10**(2):195-213. DOI: 10.15838/esc.2017.2.50.11

[44] Ferrari G, Mondejar J, Secondi L. Tourists' expenditure in Tuscany and its impact on the regional economic system. Journal of Cleaner Production. 2018;**171**:1437-1446. DOI: 10.1016/j.jclepro.2017.10.121

[45] Hjerpe E. Outdoor recreation as a sustainable export industry: A case study of the boundary waters

wilderness. Ecological Economics. 2018;**146**:60-68. DOI: 10.1016/j.ecolecon.2017.10.001

[46] Pascariu G, Ibanescu B. Determinants and implications of the tourism multiplier effect in EU economies. Towards a core-periphery pattern? Amfiteatru Economic. 2018;**20**:982-997. Available from: https://EconPapers.repec.org/RePEc:aes:amfeco:v:20:y:2018:i:s12:p:982

[47] Suresh K, Tiwari A, Uddin G, Ahmed A. Tourism, trade, and economic growth in India: A frequency-domain analysis of causality. Anatolia–International Journal of Tourism and Hospitality Research. 2018;**29**(3):319-325. DOI: 10.1080/13032917.2017.1408025

[48] Tohmo T. The economic impact of tourism in Central Finland: A regional input-output study. Tourism Review. 2018;**73**(4):521-547. DOI: 10.1108/TR-04-2017-0080

[49] Cannonier C, Burke M. The economic growth impact of tourism in small islands developing states—evidence from the Caribbean. Tourism Economics. 2019;**25**(1):85-108. DOI: 10.1177/1354816618792792

[50] Hsu P. Economic impact of wetland ecotourism: An empirical study of Taiwan's Cigu Lagoon area. Tourism Management Perspectives. 2019;**29**:31-14. DOI: 10.1016/j.tmp.2018.10.003

[51] Laterra P, Nahuelhual L, Gluch M, Sirima X, Bravo G, Monjeau A. How are jobs and ecosystem services linked at the local scale? Ecosystem Services. 2019;**35**:207-218. DOI: 10.1016/j.ecoser.2018.11.011

[52] UNWTO. UNWTO Recommendations on Tourism and Rural Development. Madrid: UNWTO; 2020. ISBN 978-92-844-2217-3

[53] Naranjo, Gómez J, Castanho R, Loures L. Evolutionary dynamics in Mediterranean landscapes: The changes in forests and semi-natural areas in the Iberian Peninsula–A study from 1990-2018. In: Management Association, editor. Research Anthology on Ecosystem Conservation and Preserving Biodiversity. USA: IGI Global; 2022. pp. 1598-1609. DOI: 10.4018/978-1-6684-5678-1.ch077

[54] Castanho R, Couto G, Naranjo J, Pimentel P, Carvalho C, Sousa Á, et al. Evolutionary dynamics in Azorean landscapes: The land-use changes in forests and semi-natural areas in the archipelago from 1990 to 2018. In: Rocha et al., editors. Trends and Applications in Information Systems and Technologies. Switzerland: Ed. Springer; 2021. pp. 244-252. AISC 1367. DOI: 10.1007/978-3-030-72651-5_24

[55] McGarigal K, Compton B, Plunkett E, DeLuca B, Grand J. Designing sustainable landscapes: Project overview. In: Designing Sustainable Landscapes Project Technical Documents 2. Report to the North Atlantic Conservation Cooperative, US Fish and Wildlife Service, Northeast Region; 2018

[56] Kato S. Greenspace conservation planning framework for urban regions based on a forest bird-habitat relationship study and the resilience thinking [thesis]. UMass Amsherst; 2010. Available from: http://scholarworks.umass.edu/open_access_dissertations/212

[57] Castanho R, Garrido J, Loures L. Environmental impact of an Iberian golf course: The case study of Guadiana golf in Badajoz, Spain. WSEAS Transactions on Environment and Development. 2021;**17**:648-658. DOI: 10.37394/232015.2021.17.63

[58] Cabezas J, Martín J, Naranjo J, Fernadez-Pozo L, Castanho R. Serviços Ecossistêmicos e Geodiversidade em Paisagens do Mediterrânneo. In: Souza-Fernandes L, Aragão A, Sá A editors. Novos Rumos do Direito Ambiental. Um Olhar Para a Geodiversidade. Campinas: Ed. UNICAMP; 2021. pp. 349-366. ISBN13: 978-65-86253-87-0

[59] Castanho R, Vulevic A, Loures L, Cabezas J, Naranjo Gómez J. Redeveloping derelict landscapes on transboundary areas–fostering cross-border cooperation as a possible solution. In: Proceedings of the 24th APDR CONGRESS; 6-7 July 2017; UBI, Covilhã-Portugal. 2017. ISBN 978-989-8780-05-8

[60] Batista T, Torre S, Fermoso J, Morais N, Avila R, Pinto C, et al. Nature-based solutions for climate change adaptation in scholar and social buildings: Concepts and pilot solutions for Portugal and Spain. In: Castanho R, Férnandez J, editors. Urban Green Spaces. London, UK: Intech Ed.; 2022. DOI: 10.5772/intechopen.105701

[61] Raposo M, Castanho R, Machado M, Castro C, Santos P, Pinto-Gomes C The relevance of vegetation series on the maintenance and sustainability of public spaces in the south-west Iberian Peninsula. In: Raposa M. Landscape Architecture: The Sense of Places, Models and Applications. London, UKIntech Ed.; 2018. pp. 254-272. ISBN 978-1-78923-713-9

[62] Rao P, Gupta J. Energy-efficient landscape design. In: Sepasgozar S, Shirowzhan S, Sargolzae S, Bienvenido-Huertas J, editors. Design of Cities and Buildings-Sustainability and Resilience in the Built Environment. London, UK: IntechOpen Ed.; 2021. DOI: 10.5772/intechopen.94904

[63] Crespo A, Velázquez J, Herráez F, Gülçin D, Ugur O'zcan A, Hernando A, et al. Territorial planning of rustic land constructions and their adaptation to climate change in the province of Málaga, Spain. Land Use Policy. 2023;**129**:1-14. DOI: 10.1016/j.landusepol.2023.106644

[64] Loures L, Naranjo J, Castanho R, Loures A. Benefits and limitations of public involvement processes in landscape redevelopment projects-learning from practice. In: Vaz E, editor. Regional Intelligence Spatial Analysis and Anthropogenic Regional Challenges in the Digital Age. Switzerland: Springer Ed.; 2020. pp. 29-48. DOI: 10.1007/978-3-030-36479-3_3

[65] Castanho R, Couto G, Pimentel P. Principles of sustainable tourism and cultural management in rural and ultra-peripheral territories: Extracting guidelines for its application in the Azores archipelago. Cultural Management: Science and Education. 2020;**1**:9-24. DOI: 10.30819/cmse.4-1.01

[66] Santos R. O regresso dos emigrantes Portugueses e o desenvolvimento do turismo em Portugal [thesis]. Universidade de Aveiro; 2013. Available from: http://hdl.handle.net/10773/10353

Chapter 4

Post-COVID-19 Impact on the Oman Tourism Market: Coping Strategies of the Hospitality Sector in Dhofar Governorate

Ali Said Akaak, Mohammed Nazeer Ahmed, Navaneetha Krishnan Rajagopal and Khalid Salim Al-Shanfari

Abstract

The COVID-19 pandemic spread rapidly worldwide. The tourism industry has suffered deep losses that have profoundly impacted attracting tourists and services offered, where many hotels are temporarily or permanently closed. This exploratory study has been conducted to study the impact of the pandemic on the hospitality sector in Oman. It contributes to the literature by providing new managerial insights and perspectives on the management of hotels during crises. Employing a case study approach supported by qualitative analysis, the study investigated the experiences of hotel managers in coping with challenges presented by the pandemic, the benefits of government support programs, employee retention, and adopting revenues and cost-saving strategies during the crisis. The study collected authentic opinions through semi-structured interviews with hotel managers in the Dhofar region (Oman). The analysis revealed that hotel managers suffered from several issues (challenges): stress, shortage of employees, financial losses (e.g., liquidity problems), closures, reopening, and difficulties adapting to change. This study emphasizes coping strategies to mitigate the onslaught of the COVID-19 crisis, which the tourism industry encountered amid the pandemic and post-pandemic. The study also provides suggestions that hotel managers, in collaboration with government officials, need to set up more support programs that will address the challenges faced by the hospitality sector, which is the crux of this research.

Keywords: COVID-19, pandemic, government commitment, coping strategy, hospitality businesses, Dhofar, Oman

1. Introduction

The COVID-19 pandemic has delivered one of the worst economic shocks in modern history. Since declaring COVID-19 as a pandemic on March 11, 2020, it has affected humans at health, educational, humanitarian, social, and economic levels [1]. Until now, its impact has paralyzed many industries and consequently, the economy, whether within the country or between countries, causing substantial financial damage that has resulted in the economies of the countries being devastated. This led to massive job losses [2] and caused great stress among stakeholders. Economically, the pandemic induced a severe economic recession due to the closure of borders between countries and territories, which disrupted the global economy as well as the performance of businesses [3].

Among the industries most seriously affected by the pandemic is the tourism and hospitality industry. Many countries around the world depend on tourism as a key economic driver or a strategic growth sector. Since the 1950s, after WWII, the industry's growth continued to increase up until the outbreak of the COVID-19 pandemic restrictions in 2020. The outbreak of COVID-19 has weakened the industry globally, leading to closures in hospitality, lockdowns, accommodation service shortages, and inflated prices. For example, holidays and business travels have been disrupted due to continuous lockdowns and international border closings. The shock to the global tourism sector could affect 45–70% of output, depending on the severity of the pandemic in that particular area. The UN World Tourism Organization (UNWTO) estimated that the world would lose 25 million jobs in economic turmoil, more than what happened during the global financial meltdown in 2008. Due to the COVID-19 crisis, the global tourism industry suffered a revenue loss of $1.3 trillion in 2020. The reason was clearly evident that the traveling plans of tourists had fallen drastically. The revenue loss is considered to be more than 11 times the loss when compared with 2009 global economic crisis [4]. The Organization for Economic Co-operation and Development (OECD) estimated an anticipated 60 to 80% decline in international tourism, while the UNWTO predicted a decline in international tourist arrivals by 58 to 78% in 2020, amounting to a loss of USD 910 to 1170 billion in international tourism receipts [5]. In addition, according to the IMF's April 2021 World Economic Outlook Report, the global economy contracted by 3.5% in 2020, a 7% drop from the 3.4% growth anticipated in October 2019 [6]. Additionally, there were unprecedented global employment losses of 114 million jobs in 2020 relative to 2019. It is estimated that the resulting reductions in travel in 2020 alone deleted US$ 4.5tr from the global tourism economy and cost millions of jobs.

In Oman, tourism represents a major economic pillar for Oman's economy and a key component of economic diversification for oil-exporting countries. In 2019, the travel and tourism sector accounted for 2.4% of GDP growth and more than 100,000 tourism-related jobs across the Sultanate, according to the NCSI [7]. However, due to the pandemic, the sector has faced unprecedented challenges. Restrictions on foreign travel have had a tremendous impact on the local economy over the last 18 months (March 2020–October 2021). In the first quarter of 2021, outbound international flights taken by Omani tourists were down 94% compared to the year before. The tourism sector suffered the largest reductions in tourist arrivals in 2020, estimated at between 70% and 80% [8]. Due to the COVID-19 crisis, fewer tourists were visiting Oman; this had directly impacted on tourism sector business. According to the statistics issued by the Ministry of Heritage and Tourism, there was a 75.1% drop in tourists in the year 2020 when compared with the year 2019 [9]. Further, according to the Tourism Under-Secretary at the Ministry of Heritage and Tourism (MOHT), the COVID-19 pandemic in the Sultanate has caused direct losses to the tourism sector that are estimated at half billion Omani Rial as of the end of September 2020 [10]. In the Sultanate of Oman, there was a steep decline in hotel occupancy by guests, both foreign and local tourists. This situation resulted into the decline of total revenues of Omani hotel businesses. According to Oman observer daily (Nov 29, 2020), "The total revenues of Omani hotels in the three-to-five-star category fell by 60.2% to RO 70.70 million until the end of October 2020 compared to RO 177.72 million for the same period in the previous year". For example, the Omani International Hotel Management Company recorded losses of 66.2 thousand Omani riyals (172.2 thousand dollars) during the first half of this year, compared to profits of OMR 1.2 m ($ 3.1 m) during the first half of 2019, according to a statement issued by the company on July 12th. Its revenues fell by 51.9% during the first half of 2020, compared to the first half of 2019, to reach approximately OMR 2.6 m (US$ 6.8 m), according to the statement [11].

This study is about investigating the COVID-19 pandemic effects and the measures that the government and hotel managers have undertaken: (i) to mitigate the impact of the pandemic on the hotel sector in Dhofar Governorate, Oman; and (ii) to identify the government actions and support requirements that have assisted in mitigating the negative impacts on the hotels. The research has raised many questions: "How did the managers of (3-, 4-, and 5-star) hotels in the Dhofar Governorate manage their hotels during the pandemic?" Sub-questions arose from this: (a) How did hotel managers perceive the impacts of the COVID-19 pandemic on the hospitality sector? (b) What are the best practices followed by the Omani government in dealing with the effects of the emerging COVID-19? (c) To what extent have COVID-19 constraints caused hotel management to adapt to the pandemic?

2. Contribution of the study

The importance of this study is that it contributes to the literature by illustrating governments' measures and aid strategies in assisting and supporting hotels during the pandemic in the under-researched hospitality sector. Following the global COVID-19 pandemic and its devastating impact on every industry, particularly the hospitality industry, there are calls to conduct theoretically driven and systematic research into the experience of hotel managers and the government's support strategies.

There are extensive studies that have investigated the impact of a crisis on tourism in different contexts, such as hospitality [12–14] or with specific reference to travel agents [15]. Other researchers have focused on the economic crisis as a form of crisis in tourism [16–19]. Although several authors analyzed the economic impacts of the COVID-19 pandemic, none of them have specifically examined the Sultanate of Oman in the context of the Dhofar Governorate and the joint actions taken to overcome the pandemic's impact.

This study is intended to serve as a reference for hospitality planning and management through extended knowledge about the key concerns and ways of coping with the changes and adjustments undertaken by the Omani government and hotel managers during the COVID-19 outbreak. In addition, it contributes to an increase in scientific knowledge about the effects of the pandemic so that the hotel sector stakeholders, such as government officials and hotel managers, can get hotel operations back to normal. Moreover, this study presents an understanding of the key trends and implications for Omani tourism and hospitality. The study will be shared with the government, concerned parties, and academic institutions to support future research in the hospitality sector and other similar sectors. It also seeks to help decision-makers respond and make evidence-based decisions. Again, this study can effectively assist in adjusting the sector development strategy, mitigating risk to the local tourism market.

According to the authors' knowledge, this is the first research to examine the influence of the COVID-19 pandemic on the hospitality sector in Oman, especially in the Dhofar Governorate during the pandemic's peak period. As a result, the authors of the study believe that the findings of this study will benefit a variety of parties, including current and potential investors, government officials, scholars, and researchers, by providing a better understanding of the relationship between the COVID-19 pandemic and hotel business in emerging destinations. The study also assists in identifying long-term trends that will be part of the process to anticipate adaptive measures for the industry moving forward. Furthermore, this study contributes to forecasting the future to set driving measures and policies for the recovery of the sector and the ways the government needs to develop a focus on safety and overcoming the challenges. The results are expected to demonstrate the Omani government's policies that have been geared toward hotel management practices as they struggle with the COVID-19 crisis. In addition, the measures and policies are expected to provide hospitality professionals with preliminary insights and recommendations to manage the effects of the COVID-19 crisis.

The remainder of the study is organized as follows. In addition to the current introduction, the study summarizes previous empirically related studies. After, the research outlines the methodology and data collection. Next, the analysis of the results and discussions focus on the pandemic's impact on hotel managers in the Dhofar Governorate. Following this, the study discusses the analysis and results. Finally, the research presents the conclusions, limitations, and future research horizons.

3. Literature review

Among various service sectors of the Omani economy, the spread of COVID-19 caused substantial damage to the Sultanate hospitality sector. According to the official reports (MOHT and NCSI), hotels account for more than 60% of all accommodations in the Sultanate, making them the backbone of Omani hospitality. The hospitality sector is at the forefront of the sectors affected, which has experienced heavy job and revenue losses. Importantly, due to the authorities' decision to prevent gatherings in public places, including gatherings on beaches and picnic areas, such as valleys, mountains, sands, waterfalls, and water springs. According to a report in the Times of Oman, government institutions, as a response to the spread of the pandemic, namely the Ministry of Tourism, Muscat Municipality, and the Ministry of Health, have released restricted guidelines for hotel staff and guests. These measures to counteract COVID-19 in hotel establishments were linked to decisions made by several organizations, namely the Ministry of Tourism, Muscat Municipality, and the Ministry of Health. The Ministry of Heritage and Tourism (MOHT) issued instructions to the hotels on the steps that need to be undertaken to minimize COVID-19 contagion risks. All hotel establishments need to limit entrances, provide transparent face masks for staff and make hand sanitizers available for the guests and staff. The restrictions also directed the hotels to keep swimming pools and gyms closed until further notice. Furthermore, the instructions also include specific guidelines on elevator use and operations at hotel kitchens. As a result, hotel groups experienced a significant reduction in sales. The already crippled tourism business in the Sultanate of Oman had to face much more severity with the cessation of the cruise ship and airline activities by the concerned authorities in Oman. This attempt with the good intention of the

Omani authorities to curb the pandemic from further spreading, had caused numerous issues and dilemmas for hotel reservations and tourist trip that were contacted prior to the pandemic [20].

According to the National Center of Statistics and Information (NCSI) [8] reports, the hotel sector, which comprises the bulk of the tourism sector's workforce, will not be able to return to its previous level very soon either. Indeed, even if activities have resumed after letting up loosening the restrictions, sanitary rules still limit the occupancy rate of establishments. In March 2020, most hotels were forced to reduce the salaries of their employees and give them leave of up to 90 days in order to reduce the proportion of financial losses incurred by the cancelation of reservations and the cessation of tourism. Moreover, to keep up their status in the market, hotels had to adapt to the 'new norms' and present the ways hotels attract the guests. As mandated by the Supreme Committee on COVID-19, all passengers arriving in Oman must undergo institutional quarantine for a minimum of seven nights. The move by the Sultanate of Oman to do the hotel bookings through the official government platform ensures that everyone abides by the quarantine rules, so as to curb the spread of the pandemic [21].

4. Hotel establishments in Oman

According to the NCIS [8], the number of hotel establishments increased in 2020 to 547 hotel establishments compared to (492) establishments in 2019, an increase of (11.2%), saving (26,733) hotel rooms and apartments in 2020 compared to (25,408) an increase of (5.2%) over the year 2019. The Governorate of Muscat had the highest number of hotel establishments, represented by (35.5%). Out of the total hotel establishments, followed by the Governorate of South Al Sharqiya, which represents (14.3%), the Governorate of Al Dakhiliyah (11.7%), then Al Batinah South Governorate (6.6%), while in both Dhofar and Al Batinah governorates, North Al Batinah and North Al Sharqiya Governorate represents 6.4%, followed by Al Buraimi Governorate with a percentage of (5.5%). In both Al-Dhahirah and Al-Wusta governorates, their share of hotel projects reached (2.7%), and the governorate of Musandam ranked last in investment in hotel projects, with a percentage of only (1.8%).

Table 1 illustrates the increase of hotel establishments in Dhofar Governorate in the last 5 years (2017–2021), according to the late statistics from the NCSI. 5-star hotels have increased from four establishments in 2017 to 8 in 2021. On the other

Hotel class	2022	2021	2020	2019	2018	2017
5-Star	8	8	8	7	6	4
4-Star	3	3	3	3	2	4
3-Star	8	5	5	4	2	2
2-Star	6	7	6	6	8	7
Others	16	15	13	13	16	16
Total	41	38	35	33	34	33

Source: NCSI [7], NCSI [8], MOHT [22].

Table 1.
Number of hotels and class of hotel—Dhofar Governorate.

Hotels guests	2019	2018	2017
Omani	102	122	115
GCC	32	57	51
Other Arabs	8	19	12
Europeans	80	253	163
Asians	41	24	33
Other	7	15	8
Total (000)	270	490	382

Source: NCSI [7], NCSI [8].

Table 2.
Number of hotels guests—Dhofar Governorate.

hand, 4-star hotels have shown a decrease of 1. Surprisingly, 3-Star hotels showed a big jump of almost fourfold from 2017. Other hotels have shown no changes.

The interesting aspect here is the number of guests at hotels in Oman increased by 22.6% during the last 6 months ending June 2021. The guests in physical number, in the year 2021 has increased to 5,22,329 compared with 425,905 in the same period of the year 2020 [23].

Table 2 illustrates that the number of hotel guests that visited Dhofar from 2017 to 2019 has increased while comparing the figures of 2017 with the figures of 2018. However, when the figures for 2019 are compared with the figures of 2018, there is a decrease in the types of guests such as Omani, GCC, other Arabs, Europeans, and others. However, only the Asian guest number is more in the year 2019 when compared with the figure of 2018. The increase in Asian numbers cannot be ruled out; perhaps, some of the Asians had traveled to Oman for employment opportunities.

However, in the first half of 2021, the figures are promising. According to Muscat Daily [9], "Number of Omani guests rose by 164 percent to 369,666 in the first half of 2021 from 139,798 Omani guests during the same period in 2020. This was followed by 74,127 Asian guests this year compared to 42,722 in the first half of 2020, comprising a rise by 73.5 percent."

As shown in **Table 2** below, tourist arrivals to the Sultanate were no better in 2021. Still, at the end of the year, from October, when the winter season starts, the hospitality sector has shown significantly better occupancy in Muscat and Dhofar hotels.

5. Employment

The tourism industry is an important source of job openings for locals; however, jobs and tourism businesses are at risk. In Oman, half of all the people working in tourism lost their jobs, especially those in the hotel sector. The UN Conference on Trade and Development report estimates a further US$1.7tr to US$2.4tr could be lost from the global tourism sector by the end of 2021. The Tourism Under-Secretary at the Ministry of Heritage and Tourism commended the support of the Omani government to the tourism sector. The Omani government gave consent to reopen the activities and provided incentives to the private sector. This effort from the Omani government during the crisis could secure the stability of jobs of not less than 15,700 people who work in the tourism sector in the Sultanate of Oman [10].

5.1 Hotel revenues

Globally, hotels of all classes (1-to-5-star) are struggling with low occupancy rates amid ongoing travel restrictions for almost 2 years. In Oman, hotels continue to reel from the historic drops in revenues and mounting losses due to the impact of the COVID-19 pandemic. The official reports from NCSI stated that the hospitality sector has faced growing financial stress as hotels face a shortage of guests and tourists amid travel restrictions. According to WAFOMAN [14]: the revenues of the Sultanate's 3 to 5-star hotels decreased from OMR 24 million in April 2019 to OMR 1 m in April 2020. The month of April is the first whole month for hotels to spend after the Sultanate stopped all domestic and international flights on March 29th. The official data issued by the National Center for Statistics and Information show the size of the actual impact of the closure imposed by the spread of COVID-19 on the hotel sector in the Sultanate. According to data issued by the official center of the Sultanate, the number of guests in this category of hotels reached only 12,000 in April 2020, compared to nearly 200,000 guests in April of last year, and the occupancy rate decreased from 63% to only 16% [14].

Table 3 shows that the sector also requires a direct link with the customers (hotel guests). It has been affected by the decline in the number of visits to the Sultanate and the collapse in demand by tourists and business travelers due to the closure of borders. Total revenues of Omani hotels—within 3- to 5-star categories—further plunged by 35.8% to OMRO 37 m in the first half of 2021 compared to OMR 57.67 m recorded in the same period of 2020 [23].

The disruption caused by the pandemic and global economic decline, especially in the hospitality sector, require a fast-moving transition and adoption strategy, a recovery strategy for the hospitality sector. The undefined persistence of the pandemic duration raises anxiety about the ability to recover from this situation, since hotel businesses from the hospitality sector are known to be relatively more susceptible to

Governorate	2019	2018	2017
Muscat	53.1	54.8	55.3
Dhofar	41.6	46	49.1
Musandam	22.7	25.7	35.2
Al Buraimi	42.1	41.9	42.6
Ad Dakhiliyah	14	14.1	22
Al Batinah North	44.4	44.3	44.3
Al Batinah South	41.2	41.8	42.5
Ash Sharqiyah South	37.9	37.9	34.3
Ash Sharqiyah North	13.7	13.1	14.4
Adh Dhahirah	13.5	13.3	26.1
Al Wusta	34.4	30.7	40
Occupancy Rates	37.4	38.4	45.2

Source: NCSI [7], NCSI [8].

Table 3.
Occupancy rates by governorate.

the economic complications arising from a crisis. According to NCSI [8], many hotels decreased their occupancy almost to zero during the first 4 months of this year due to the closure procedures, which exposed them to losing tens of millions of Omani rials despite government support. Moreover, tourists' perceptions, future expectations, and spending are seriously threatened by a high level of uncertainty. Official data indicate a decrease in the number of visitors to the Sultanate from 289,000 in April 2019 to only 1000 when the Sultanate is preparing for two tourist seasons, the first of which begins in the Dhofar Governorate in the south of the Sultanate, which begins in June and continues until September of each year and attracts mainly visitors from the Gulf Cooperation Council (GCC) countries [14].

As a response to contain COVID-19, the Omani government has implemented preventive, restrictive, consecutive, and prolonged measures to address the pandemic and its fallout. As a result, strict containment measures helped curb the first wave and were gradually lifted from June 2020 onward. Travel slowed again as new COVID-19 case numbers began to climb toward the end of the summer and early fall of 2020, leading to a severe spike in October/November. This led to the introduction of stronger restrictions in November 2020 and a return to the COVID-19 Response Phase in the Sultanate of Oman. However, an expert panel from the United Nations World Tourism Organisation expected travel levels to the Middle East to get back to pre-pandemic levels by 2023 or 2024. Moreover, sales in the hotel services sub-sector declined by 60% (approximately). This was due to business closures across the country caused by the new regulations introduced to limit the spread of COVID-19.

As reported in many reputed newspapers in Oman, the overall occupancy rates in Oman's 3- to 5-star category hotels still remain very low. For instance, referring to the financial figures of Salalah Beach Resort Company, a luxury 5-star resort that offers a beachfront location in Dhofar, reported a net loss of more than RO475,000 for the first half of 2021 against a net loss of RO303,818 in the same period of 2020 [23].

6. Road to recovery: oman's government responses to COVID-19

Like many countries around the world, the pandemic has caused massive economic disruptions in the Sultanate through simultaneous shocks, including a drop in domestic and external demand, lower oil prices, disruption of trade and global value chains, lower consumer and investor confidence, and tightening financial conditions. Such challenges have forced governments to make quick decisions, manage crisis coordination, and implement strict measures to protect vulnerable communities.

As a response to contain COVID-19, the Omani government, has developed an 'Economic Stimulus Plan' to support the efforts to recover the impacts and repercussions of the COVID-19 on the national economy. To support tourism businesses, the plan includes developing robust policies and corporate plans, including hotels, and providing a set of procedures and initiatives designed for the current situation, enhancing the performance of economic activities, and attracting foreign investment [24]. The government came up with interventions to help tourism businesses such as hotels brace themselves against the prolonged and pronounced impacts of COVID-19. The Omani government has also developed recovery scenarios in collaboration with partners to identify the framework of multiple phases of recovery, which align with the global opening to travel. These scenarios are developed to set a tourism recovery plan to boost Omani's domestic and inbound international tourism following the pandemic. As well as improving transport links, restructuring the organizations that manage Oman tourism, and investing in infrastructure, it is focusing on domestic marketing to promote the Sultanate as a holiday destination for GCC citizens. After the severe setback due to COVID-19 worldwide, tourists are reluctant to travel overseas. Hence, they are forced to depend on touring around within their own countries. This is basically a conservative and precautionary approach followed by tourists, which may benefit internal tourism [25].

The Omani government has taken initiatives to support the tourism industry. Measures include specific credit lines for tourism enterprises introduced by the Central Bank of Oman, exemptions from tourism levies and other fees, and debt rescheduling. Particular attention has also been given to the tourism sector. The Ministry of Finance decided to postpone the payment of tourism fees imposed on hotels—amounting to 4%—until the first quarter of 2021 to mitigate the economic impact on the sector. In addition, coordination between the Ministry of Tourism and the Royal Oman Police extends the validity of unused tourist visas until March 2021 [14]. In an effort to support financially to the affected sectors due to COVID-19, the Omani government through the Central Bank of Oman has issued a number of directives to all banks and finance leasing companies (FLCs), including the announcement of a stimulus package of procedures and incentives aimed at containing the economic impact of the COVID-19 pandemic on the lives of individuals and institutional borrowers [26]. Additionally, in a bid to help small and medium enterprises, the Omani government has also adopted several measures including manpower support, deferred payments and also interest-free loans from the Development Bank [27]. To curb the menace of COVID-19, the Ministry of Tourism, Oman has taken extensive measures to curb COVID-19. The extensive measures include high-level training for all operational staff in virus protection measures to control the safety of all. Further, all guests will be provided with mask and disinfectant upon their arrival [28–30].

The government has introduced extensive economic support packages and programs that have helped tourism businesses brace themselves against the prolonged impacts of COVID-19. To assists financially to businesses which are suffering from the economic downturn, the Central Bank of Oman (CBO) has unveiled financial incentives of RO 8 billion in the form of additional liquidity [28–30]. Nevertheless, these costs and future expenditures cannot be extended indefinitely. According to WAFOMAN [31], the Omani Ministry of Finance has approved a proposal to postpone the payment of tourism fees on hotel establishments—submitted by the Ministry of Tourism—until the first quarter of the year 2021. Former Omani Minister of Tourism stated - during his participation in the weekly conference on the developments of the Coronavirus in the Sultanate—that the Ministry of Finance agreed to postpone these 4% fees payable during the period from December 1, 2019, to the end of August 2020, provided that the payment be made in the first quarter of next year after recovery [31]. The former minister explained that the procedure aims to enable hotel establishments to benefit from the liquidity provided by these fees—which are supplied to the state treasury through the Ministry of Tourism—to run the work of hotel and tourist facilities.

7. Research methodology

This study investigates the COVID-19 impacts on the hospitality section in Oman for the period March 2020–October 2021. It also highlights the devastating effects of the pandemic on the hospitality sector and hotel managers. This research is based on a qualitative approach to achieve an in-depth understanding [32] of hotel management experience during the pandemic. The case study method was employed [33] with the hotel managers as the unit of analysis.

Since the study is studying the cases of hotel managers in Dhofar, it is believed that the findings' transferability to comparable contexts will be beneficial for both research and practice. The methodology used is of a qualitative nature, based on conducting semi-structured interviews to study the impact of the pandemic on the hospitality industry with a specific emphasis on employment and revenues. Through semi-structured interviews, the study collected authentic opinions on the experiences of (3-4-5 star) hotels (international or local hotel chains) from Dhofar Governorate after the COVID-19 pandemic. Using this approach, the study sought to examine the perceptions, feelings, and emotions of hotel managers in the context of the impacts of a financial crisis.

The targeted population included 3-, 4-, and 5-star hotel managers in Dhofar Governorate (Oman) who operate in the hospitality sector. The sampling approach was as follows. From the official annual reports of the Ministry of Heritage and Tourism (MOHT), published in January 2021, the study was able to obtain a list of hotels in the governorate, numbering 19, see **Table 4**. Then, for each hotel, the study consulted the database of the MOHT to determine the type of business and identity

Hotel name	Class	No. of hotels
Crowne Plaza Resort Salalah	5-Star	8
Whynadm Garden Mirbat Resort	5-Star	
Salalah Rotana Resort	5-Star	
Al Fanar Hotel – Salalah Beach	5-Star	
Millelumen Salalah Resort	5-Star	
Hilton Salalah Hotel	5-Star	
Alila Hinu Bay – Mirbat	5-Star	
Al Baleed Resort Salalah by Anantara	5-Star	
Juweira Boutique Hotel	4-Star	4
Salalah Gardens Hotel	4-Star	
Belad Bont Resort	4-Star	
Tulip Inn Majan Hotel	4-Star	
Muscat International Hotel Plaza Salalah	3-Star	7
InterCityHotel Salalah	3-Star	
Hamadan Plaza Hotel	3-Star	
Atana Stay Salalah Hotel	3-Star	
Bristol Hotel Salalah	3-Star	
The Plaza Hotel & Resort	3-Star	
Salalah International Hotel	3-Star	
Total	19	

Source: Filed Study.

Table 4.
Interviewed hotel managers in Dhofar Governorate.

of its owner. This allowed them to retain only 3 to 5-star hotels, and obtain a list of their managers as well as their addresses. Thereafter, we proceeded with purposeful sampling [32] by selecting managers based on their lived experiences and knowledge during the COVID-19 crisis [34]. The purposeful sampling method is intended to capture the characteristics of both commonalities and differences that are necessary for producing new knowledge [35]. As Bernard [36] and Spradley [37] explain, the validity and reliability of the sample are based on the availability, willingness, and ability to participate, communicate, and share experiences in a reflective manner. Qualitative methods are intended to achieve a deeper understanding [32], while sampling methods are intended to maximize efficiency and validity [38]. Miles and Huberman [39] place primary emphasis on saturation. Therefore, the study limited the interviews to 19 managers in person.

All the participants have responded to us with their written consent. The interviews were conducted face-to-face which assisted the participants to share their experiences faced by them during the COVID-19 crisis. Using the appropriate software NVivo, the responses were coded. In the process three main themes are developed. The first theme explored the experiences of hotel managers during COVID-19, the second theme highlighted the innovative packages to overcome strategic crisis management and the final theme was the lessons learned in the wake of COVID-19 crisis. Equally important, this study highlights the use of secondary data, where collecting data and information came from official sources such as NCSI and articles (JSTOR and Scopus database) about Oman from March 2020 to December 2021.

8. Results and discussion

This study analyses the results of 19 in-depth interviews with managers representing different hotel classes (3-, 4-, and 5-star) in the Dhofar Governorate, Oman. These hotel managers have experienced the profound impact of the pandemic on their hotels' business. They expressed their concern regarding how the COVID-19 pandemic has greatly affected hotels and even threatened the prosperity and existence of their businesses. A 5-star hotel manager spoke about the impact of the crisis on tourism, saying: "There is no longer a field for tourism, due to the suspension of air traffic, the closure of airports, and the fear of individuals to move with the spread of the epidemic."

8.1 Hotels financial losses

The findings have highlighted that the hotel sector has experienced financial losses in the wake of the COVID-19-induced financial crisis. Until now, the sector has suffered from the pandemic's financial impacts, including financing difficulties, recession, and decreased revenues. Hotels found it challenging to gain income due to the drop in the number of guests. The number of guests reached an extreme low during the pandemic. In addition, many hotel projects collapsed dramatically or were canceled in the wake of COVID-19, resulting in a 70% decrease in total guest volume, according to a 4-star hotel manager. According to some hotel managers, the most significant burden in this sector falls on big hotels (4- and 5-star). He argued that since the onset of the crisis, hotels had experienced a sharp and dramatic decline in overall service levels. The severity of this decline far exceeds the declines seen in more established hotel markets. Thus, the findings confirmed that most hotels have insufficient financial resources and have recently spent almost all their reserves. Without the ability to generate revenue, hotels had to continue to assume the costs of their outstanding expenses. The study revealed that most hotels have insufficient financial resources.

All service sectors are connected, and therefore, the slowdown in hotel services has caused a deviation in incomes to the other business sectors due to decreased sales revenues. Global containment measures and travel restrictions implemented by several countries also disrupted supply, production, traveling, and consumption, resulting in massive revenue losses. According to some managers, the pandemic made strategic impacts on the global supply and demand chain. A 5-star hotel manager shared that hotels faced supply chain issues as workers were confined to their homes, while operations were curtailed in some industries. Partial or total business closures, emphasis on facility hygiene, and protective measures to comply with new measures and restrictions also affected supply chains. A 3-star hotel manager commented, "At the beginning of the pandemic, managers had many difficulties adapting to this new reality. They had to set up different arrangements while complying with the authorities' guidelines, which changed their lifestyle and managerial culture."

The interviewees and managers argued about the restructuring of loans to investors in the hotel sector, financing policies, encouraging banks to invest in the tourism sector, and providing a package of exemptions such as exemption from the tourism tax represented by 4% and the municipality tax by 5%, as well as exemption from fees for tourist licenses at these facilities. In addition to this, exemption from tourism and municipal taxes is represented in restaurants, cafes, event halls, gymnasiums and other services in these facilities for a period of 2 years, subject to renewal. Further, granting companies and tourist offices an opportunity to replace tourist visas and review the electricity tariff for these facilities to be 20 baisas for a period of 2 years, subject to extension. Accordingly, the circumstances with some other matters related to the workforce from extending the previously offered incentives and providing another incentive package in support of the tourism sector, represented in suspending the requirement of Omanization rates and exempting workers in the sector from renewal fees during this period so that establishments can adapt to this stage and its repercussions. The interviewees stressed the great importance of paying attention to this sector and supporting it in various ways, as it is one of the most affected sectors at global and local levels, calling for rapid intervention by those concerned to support them to survive this pandemic with minimal losses.

8.2 Hotels health measures

The study confirmed that COVID-19 still has a negative impact on hotel performance, which inversely affects offered services and tourists' decisions to visit át-risk' tourist destinations. Most, if not all, hotels at different classes had introduced stringent sanitary protocols within their organizations (deep cleaning, easy access to gel and wipes, temperature screening, social distancing, face masks and gloves, online/touchless registration, and payments, upgraded ventilation systems, and individual single-use meal boxes). In addition, luxury hotels during the pandemic shifted from traditional guest protocols to new ones that suited the current situation:

- Incorporating permanent video conferencing technology in meeting rooms.
- Re-mapping of meeting capacities to respect social distancing.

- Implemented 12- to 24-hour room shutdown between guests.
- Eliminated all buffet service.
- In some cases, hotels have been repurposed into COVID-19 recovery and/or quarantine centers, long-term care facilities.
- Reconsidering contracts for future bookings.

8.3 Government support programs

Regarding the role of the government in boosting the sector, several governments worldwide have provided financial support and grants to tourism businesses to mitigate the negative impacts of the crisis, ensure the survival of their operations, and limit the loss of employment opportunities. The results of the study showed that government support for hotels were and still are the most crucial factors that shaped the immediate adjustments of operational activities in response to COVID-19. These support programs have come in different packages to bring back tourists and sustain the hospitality sector during the COVID-19 pandemic. The government continued feeding the media with information related to strategies implemented toward reviving the tourism and hospitality sector, which will build the confidence of the tourists and businesses during the pandemic period. According to an interviewee with a 4-Star hotel manager, the government set out the different COVID-19-related public rescue strategies for the tourism and hospitality sector at the business level (for tourist companies).

Regarding the benefit of government support programs, some hotel managers assured that the government support packages had helped them better manage the pandemic's impact. They assured that the government was mainly available to support the companies financially. A hotel manager empathized that government assistance has been essential in successfully managing the crisis. This has allowed local hotel businesses to have cash flow and better manage their recovery. This was reflected in the words of a hotel manager, who emphasized, "the assistance was greatly appreciated and needed; without these grants, I would not have been able to reopen." However, some hotel managers were not satisfied with the packages offered. They argued that despite the government providing support packages for the tourism sector, these packages stopped at the end of last July 2022, and hotels still suffer from a significant decline in occupancy. Again, they found the government support programs were not at a high level of expectations, especially at the beginning of the pandemic, and even later, when tourism was restored.

8.4 Managing hotels during the pandemic

The COVID-19-induced financial crisis put hotel managers under substantial stress. The interviewees said they had experienced tremendous stress and were still experiencing it. It is certain that at the beginning of the crisis, it was very stressful because nobody knew where they were going with the financial crisis caused by COVID-19. As a response to the severe impact of the pandemic on the hotel sector, hotels had displayed a strong managerial orientation to deal with the challenges as a means of managing the COVID-19 crisis. Some 4- and 5-star hotel managers assured that some hotels, for instance, international chains, had to adopt drastic measures to stall the decision to close hotels. Still, most hotels were not spared from this, as cutting costs was not enough to cope with this terrible crisis. Other hotel managers believed that under the new work environment imposed by the pandemic, they had to develop adaptive strategies by being more aware of other opportunities. One of the interviewees urged that hotels must reorganize their services to meet consumer needs better and focus on adjusting the service prices. Also, hotel managers had to share business conditions to anticipate changes. This allows hotels to adapt quickly to focus on new trends and, thus, better manage the changes induced by the COVID-19 crisis. Thus, in the long run, cooperation between the hotel and its employees is needed by focusing on both prevention and response to significant changes. According to the interviewees, they stated that most hotels took immediate actions that encompass deferring investments, laying off workers, reducing labor costs and operating expenses, negotiating contract terms and conditions, and reducing stocks. Moreover, it is crucial to be responsive and clarify the development of the situation through different means of communication. For example, several governments recommended that employees in all sectors telecommute during the pandemic and provide remote service delivery.

In response to the implications of the pandemic, some managers emphasized the importance of maintaining a contingency financial account. This is particularly to cover certain contingencies until the help from the government in the form of aid and subsidies arrives. However, some managers felt that having a contingency financial account will be an obstacle to obtain financial subsidies from the government. The study identifies that managers with several years of managerial experience are more likely to handle the crisis and its aftermath effectively. Also, the study observed that experienced managers are having the capability to adapt to change rather than resisting the change.

The extent of the readiness of these facilities to receive tourists during the coming period, hotel managers shared that they have applied for a package of tax exemptions to face the repercussions of the COVID-19 crisis. They added that they have addressed many proposals and visions that could have reduced the economic losses incurred by the hotel sector during the current period due to the pandemic, praising the importance of the Ministry of Heritage and Tourism for adopting their proposals and views of the supreme committee for dealing with the pandemic. Other interviewees stated that they had strategically opted to stay in business to avoid losing their skilled workforce. For example, a 3-star hotel manager said that he kept the hotel open just for takeaway orders to ensure he could keep his staff. He mentioned, "One of the choices made was to keep my staff, especially my good employees." On the positive side, the pandemic-induced crisis allowed hotel managers to leverage the pandemic-induced closure for refurbishments and administrative reorganization. Some hotel managers mentioned that they "decided to renovate the entire hotel such as Hilton Salalah to cope with the guests' needs."

9. Back to normal life

Successful cooperation between the government and concerned parties in the tourism sector, including civil society is a key factor in an effective response to crises. The pandemic underlined the need to accelerate public administration reform to create a more professional, efficient, effective, and transparent public sector that focuses on the satisfaction of society. Also, the competition between tourist destinations in the Gulf region improves the quality of tourism services and facilities, diversification of the tourism product, and the achievement of cost-efficiency in prices highly demanded. The finding revealed that the pandemic may impact the international tourism. Further, tourists may prefer internal tourism as a precautionary measure. But it looks increasingly unlikely, once the pandemic is over the Omanis and the foreign tourists will eager to travel as similar to pre-pandemic. According to the interviewees, it is estimated that there was a 60% reduction in domestic holidays during 2020. It also predicts that 2021's domestic tourism will remain well below that of 2019. Rather than a boom in domestic trips, many are not going on holiday at all. According to NCIS [8], the revenues of the hotel sector were slashed by more than half in 2020. Interviewed hotel managers estimated that the demand for hotels is not expected to return to pre-COVID levels until 2023.

Regarding the contribution of tourism to the economic development of the governorate, one of the interviewees commented, "What we saw in the Dhofar festival of this year's activities, the increase in the number of visitors and the complete readiness to receive tourists from all countries in the fall, was indicative of a strong and effective role of tourism in supporting economic development, and the Ministry of Tourism's interest in supporting small Omani institutions and youth during this period contributed to the success of this wonderful activity." The findings revealed that the challenges that hinder the development of the tourism sector in Dhofar the most, is the need to complete the infrastructure in some tourist sites, which the government is making efforts to strengthen, as well as the need to achieve a balance between the development of the tourism sector to reach the desired goals and at the same time, preserve the natural environment and various resources in addition to preserving the cultural and social environment of the Omani society, its customs, and traditions.

10. Coping strategies of Oman tourism: a response to the lessons learned during the pandemic

All over the world tourist choose for tourism for entertainment and exploration. It involves traveling and visiting different sites outside of their usual place of abode. The socioeconomic and political changes in the world affect the tourism industry and the growth of the travel and tourism industry has been marked by its growth after globalization and related changes around the globe. The tourism industry is promoted internationally by all the countries. It brings growth and wealth to both developing and developed countries. The tourism-oriented growth history of nations is an example of its importance as a growth agent. Conversely, tourism creates pollution and threatens indigenous culture, however, due to its heavy contribution to economic growth, nations, especially developing nations, promote tourism. Recently, the tourism phase has changed as the policymakers gave prominence to environmental protection and committed more to sustainability. Conventional mass tourism drifted to ecotourism after the world acknowledge the threat of climate change.

11. Post-COVID-19 strategies required in the tourism sector

1. *Harmless tourism environment strategy*

After the COVID-19 pandemic, the awareness of medical issues has changed the mind of the tourists who elect to travel. As a result of this awareness, there is an urge to develop a harmless tourism environment. This should be created to entice tourists and includes the facets of harmless tourism including hygiene accommodation, and ensuring disease prevention. Tourism businesses need to design strategies to create awareness and educate the tourists about illness preclusion. This strategy would help organizations to avoid tourists continuing with the phobia of illnesses or infections.

2. *Digitalization Strategy in tourism*

It is necessary to advance digitization in the tourism sector. The database has to be developed and then the strategy to go digital should be implemented. Tourism businesses need to store the travelers data in a safe and secure manner. This may lead to advanced database systems which are developed in all countries that wish to take advantage of coordinated benefits among international players in the tourism sector. This will help the tourism businesses around the world in bringing innovative terminus management.

3. *Strategy for conservation and development of human resources in tourism*

The tourism industry is at present facing the absence of a smart workforce as many people have shifted to other jobs during the last few years because of the pandemic and unstable situation. Now businesses need to focus on a human resource retention strategy by providing practical skill exercises conducted by experienced professionals as this will conserve and develop the human resources in the tourism sector.

4. *Green tourism strategy*

There is a need after the pandemic to develop green tourism. Tourists became more interested in ecotourism. There are people in the world who like to see low-carbon places and eco-friendly destinations.

Yilmaz et al. [40] created questionnaires that were completed by 395 locals in order to obtain data on the impact of regional development theories. They discovered that ecotourism was a driving force in regional development and that it had a vital role in the sociocultural and economic development of Sapanca, a town in Turkey, based on their study of tourism development plans developed in Turkey under the new regionalism theory.

5. *Future-ready strategy*

The turmoil caused by COVID-19 has resulted in declining revenue for many tourism businesses. Therefore, tourism management should focus on technology and digitization as a part of a future-ready strategy in response to overcome the obstacles that not only may be caused by new outbreaks and virus variants but also by competitive businesses; post-COVID-19, tourism businesses turned aggressively competitive.

The marketing section personnel of the tourism business should focus their attention on relevant and updated technology applications for targeting the specific audience through appropriate digital advertisements, that is, through enhanced pixels on website pages because the correct message properly targeted will prove increased viewer engagement [41].

In addition to this, appropriate information about hygiene and qualitative services will overcome the apprehensions of tourists and this will enhance the business prospects. Further, organizations are expected to invest in digitalization, that is, a higher use of automation and contactless payments as this is expected to speed up the business [41]. For increased automation artificial intelligence (AI) can be utilized by the tourism businesses assisting tourists in booking flights and hotel services. AI not only saves businesses and tourists time and money it also reduces the errors in the process.

6. *Sustainability strategy*

Understanding customers'/tourists sentiments and conditions are essential for them to purchase travel and tourism services; this approach will pave the way to ensure tourism business sustainability post-COVID-19. For sustainability, it is essential to examine the perceptions of customers/tourists on their travel and tourism expenditure. This attempt would serve as a point of reflection for destination marketers and crisis managers attempting to recover from any crisis [42].

12. Conclusion

The outbreak of COVID-19 has severely affected the tourism and travel market in developed and developing countries. The pandemic has had sociological and socio-economic implications. Despite all the challenges, the outlook for 2021 and 2022 is for considerable recovery across the Arabian Gulf. The study assisted in presenting how the Omani government action in adopting strategies and ways (measures) to mitigate the influence which could affect the sector's future. Today's governments must remain alert and adaptive to unforeseen events, such as external crises, which create increased uncertainty among their workforce and pose immediate threats to the tourism businesses performance and viability. The pandemic has imposed a severe challenge on the tourism business, as businesses are forced to navigate the unprecedented situation by finding new solutions to the challenges across many areas of their operations. This study provides valuable intelligence on the impacts of COVID-19 on the hospitality sector and the government bailout strategy to assure hotels' existence in the Omani tourism market. The study also showed the importance of government support for tourism businesses during pandemics, but specific problems persisted.

Moreover, the study suggests that hotel management should avoid hasty decisions without considering their impact and consequences on hotels. Notably, a reasonable time to adjust to the measures taken must be observed in similar future cases. The study also suggests that hotel managers, in collaboration with government officials, must set up more support programs to address the challenges. Hotel managers are encouraged to build on a culture of financial precaution by maintaining contingency financial accounts. They should also practice maintaining their employees, as human capital is essential to the business's success. Finally, the study suggests that hotel managers should receive specific training in resilience and change management during crises. The study also suggests strategies that are to be adopted by tourism businesses post-COVID-19 to garner richer benefits for the tourism business.

13. Limitations and future research

This study is subject to some limitations and, therefore, delivers some opportunities for further research. Note that when the interviews were conducted, the crisis was almost over. The interviews were conducted only with those hotels that continued to operate during the crisis. Unfortunately, the study could not interview hotel managers who ceased operations because of the pandemic. Their opinions or contributions could have been very useful in enriching the results. Although the study evidence was gathered from hotel managers in Dhofar, which assisted to define key areas that affect the hospitality sector, still, there is a need to continually gather robust data on a national basis. This evidence is vital to represent the hospitality sector in these quickly changing and uncertain times. Future research should be conducted after the end of the pandemic and include hotels across Oman and those that resumed operations after being closed for an indefinite period.

Conflict of interest

The authors declare no conflict of interest.

Funding

The author(s) received no financial support for the research, authorship, and/or publication of this chapter.

Author details

Ali Said Akaak*, Mohammed Nazeer Ahmed, Navaneetha Krishnan Rajagopal and Khalid Salim Al-Shanfari
College of Economics and Business Administration, University of Technology and Applied Sciences, Salalah, Oman

*Address all correspondence to: a.akaak@sct.edu.om

References

[1] Messabia N, Fomi P-R, Kooli C. Managing restaurants during the COVID-19 crisis: Innovating to survive and prosper. Journal of Innovation & Knowledge. 2022;7(4):100234

[2] Ratten V. Coronavirus (COVID-19) and entrepreneurship: Changing life and work landscape. Journal of Small Business & Entrepreneurship. 2020;**32**(5):503-516. DOI: 10.1080/08276331.2020.1790167

[3] Kooli C, Son ML. Impact of COVID-19 on mergers, acquisitions and corporate restructurings. Businesses. 2021;**1**(2):102-114

[4] ET HR World. 100-120 million direct tourism jobs at risk. 2021. Available from: https://hr.economictimes.indiatimes.com/news/industry/100-120-million-direct-tourism-jobs-at-risk-wto/80581577 [Accessed 27 May 2023]

[5] OECD. COVID-19 Crisis Response in MENA Countries. 2020. Retrieved from: https://www.oecd.org/coronavirus/policy-responses/covid-19-crisis-response-in-mena-countries-4b366396/

[6] IMF. World economic outlook report. 2020. Retrieved from: https://www.imf.org/external/pubs/ft/ar/2020/eng/

[7] NCSI. Statistical Bulletin 2019. Vol. 31. Muscat: National Center for Statistics and Information (NCSI); 2020

[8] NCSI. Tourism Statistical Year Book 2020. Muscat: National Center for Statistics and Information (NCSI); 2021

[9] MuscatDaily. Hotel revenues in Oman rise over 120 percent: NCSI. Muscat Daily; 9 Aug 2021. Retrieved from: https://www.muscatdaily.com/2022/08/09/revenues-for-hotels-rated-from-3-to-5-stars-in-oman-witnessed-a-remarkable-increase-of-121-4-per-cent-until-the-end-of-june-2022-reaching-ro83360000-compared-to-the-figures-of-the-same-period-in-2/

[10] Oman-Daily-Observer. Tourism Sector in Oman Loses half a Billion Rials Due to Impact of Coronavirus. Muscat, Oman: ZAWYA; 13 Dec 2020. Retrieved from: https://www.zawya.com/en/business/tourism-sector-in-oman-loses-half-a-billion-rials-due-to-impact-of-coronavirus-go37cawq

[11] Khaleej.Online. Despite Lifting the Closure, Gulf Hotels Suffer from the Consequences of "Corona". Dubai, UAE: Al Khaleej Online; 18 Sep 2020. Retrieved from: http://khaleej.online/kkRKYJ

[12] Chien GC, Law R. The impact of the severe acute respiratory syndrome on hotels: A case study of Hong Kong. International Journal of Hospitality Management. 2003;**22**(3):327-332

[13] Rittichainuwat BN. Tourists' and tourism suppliers' perceptions toward crisis management on tsunami. Tourism Management. 2013;**34**:112-121

[14] WAFOMAN. Only OMR 1m Was the Revenue of the Sultanate's Hotels in April 2020. Muscat, Oman: WAF News Agency; 2020b. Retrieved from: https://wafoman.com/2020/06/15/%D9%85%D9%84%D9%8A%D9%88%D9%86-%D8%B1%D9%8A%D8%A7%D9%84-%D9%81%D9%82%D8%B7-%D8%A5%D9%8A%D8%B1%D8%A7%D8%AF%D8%A7%D8%AA-%D9%81%D9%86-%D8%A7%D8%AF%D9%82-%D8%A7%D9%84%D8%B3%D9%84%D8%B7%D9%86%D8%A9-%D9%81/

[15] Perl Y, Israeli AA. Crisis management in the travel agency sector: A case study. Journal of Vacation Marketing. 2011;**17**(2):115-125

[16] Corbet S, Larkin C, Lucey B. The contagion effects of the COVID-19 pandemic: Evidence from gold and cryptocurrencies. Finance Research Letters. 2020;**35**:101554

[17] Khalid U, Okafor LE, Burzynska K. Does the size of the tourism sector influence the economic policy response to the COVID-19 pandemic? Current Issues in Tourism. 2021;**24**(19):2801-2820

[18] Papatheodorou A, Pappas N. Greek Tourism, Economic Crisis and Political Turmoil, between Scylla and Charybdis. Tourism and Political Change. Oxford: Goodfellow Publishers; 2017. p. 171

[19] Zopiatis A, Savva CS, Lambertides N, McAleer M. Tourism stocks in times of crisis: An econometric investigation of unexpected nonmacroeconomic factors. Journal of Travel Research. 2019;**58**(3):459-479

[20] Al-Hadhrami M. Tourism contractual obligations in light of the pandemic in the Sultanate of Oman. 2020. Available from: https://www.tamimi.com/law-update-articles/tourism-contractual-obligations-in-light-of-the-pandemic-in-the-sultanate-of-oman/ [Accessed 27 May 2023]

[21] Raajeshwari A. Oman: Sahala platform for institutional quarantine bookings in operation tomorrow. Gulf News. 2021. Available from: https://gulfnews.com/world/gulf/oman/oman-sahala-platform-for-institutional-quarantine-bookings-in-operation-tomorrow-1.78170669 [Accessed 27 May 2023]

[22] MOHT. Names of Hotel Establishments and Number of Rooms Until the End of 2022. 2022. Retrieved from Muscat: Ministry of Heritage and Tourism: https://mht.gov.om/Pages/Media/DownloadCenter/statistics.aspx

[23] MuscatDaily. Oman's hotel industry continues to reel from pandemic impact. Muscat Daily. 28 Jul 2020. Retrieved from: https://www.muscatdaily.com/2021/07/28/omans-hotel-industry-continues-to-reel-from-pandemic-impact/

[24] KPMG. Tax Alert: Oman Announces an Economic Stimulus Plan. Muscat, Oman: KPMG Lower Gulf Limited; 15 Mar 2021. Retrieved from: https://home.kpmg/om/en/home/insights/2021/03/tax-alert-oman-announces-an-economic-stimulus-plan.html

[25] OmanObserver. Tourism is back but needs a long-term vision, say experts. Oman Daily Observer. 24 Jul 2022. Retrieved from: https://www.omanobserver.om/article/1122692/oman/tourism-is-back-but-needs-a-long-term-vision-say-experts

[26] Central Bank of Oman. CBO announces directives on bank loan repayments. 2021. Available from: https://cbo.gov.om/news/277 [Accessed 27 May 2023]

[27] OmanObserver. Bank Muscat Annoucnces RO 1 Mllion to Support. 5 Oct 2021. Retrieved from: https://www.omanobserver.om/article/1107711/oman/community/bank-muscatannouncesro-1-milliontosupport

[28] Oman Observer. Ensure safety of all domestic tourists, hotel guests: Govt. Oman observer. 2020a. Available from: https://www.omanobserver.om/article/12800/Main/ensure-safety-of-all-domestic-tourists-hotel-guests-govt

[29] Oman Observer. 647K guests in Sultanate's hotels till October, 2020. 29 Nov 2020b. Available from: https://www.omanobserver.om/article/7323/Local/647k-guests-in-sultanates-hotels-till-october-2020 [Accessed 27 May 2023]

[30] Oman Observer. Central Bank of Oman unveils RO 8 bn stimulus package to fight downturn. 2020c. Available from: https://www.omanobserver.om/article/14624/CORONAVIRUS/central-bank-of-oman-unveils-ro-8-bn-stimulus-package-to-fight-downturn [Accessed 27 May 2023]

[31] WAFOMAN. Oman Adopts a Package of Facilities to Support the Tourism Sector. Muscat, Oman: WAF News Agency; 2020a. Retrieved from: http://wafoman.com/2020/06/11/%d8%b3%d9%84%d8%b7%d9%86%d8%a9-%d8%b9%d9%85%d8%a7%d9%86-%d8%aa%d8%b9%d8%aa%d9%85%d8%af-%d8%ad%d8%b2%d9%85%d8%a9-%d8%aa%d8%b3%d9%87%d9%8a%d9%84%d8%a7%d8%aa-%d9%84%d8%af%d8%b9%d9%85-%d9%82%d8%b7%d8%a7/

[32] Patton MQ. Qualitative Research & Evaluation Methods: Integrating Theory and Practice. 4th ed. Thousand Oaks: SAGE Publications; 2014

[33] Yin R. Applications of Case Study Research. 3rd ed. Thousand Oak, CA: SAGE; 2011

[34] Creswell JW, Clark VP. Designing and Conducting Mixed Methods Research. 3rd ed. Washington DC: Sage Publications; 2017

[35] Palinkas LA, Horwitz SM, Green CA, Wisdom JP, Duan N, Hoagwood K. Purposeful sampling for qualitative data collection and analysis in mixed method implementation research. Administration and Policy in Mental Health and Mental Health Services Research. 2015;**42**(5):533-544

[36] Bernard HR. Research Methods in Anthropology: Qualitative and Quantitative Approaches. Lanham, Maryland: Rowman & Littlefield; 2017

[37] Spradley JP. The Ethnographic Interview. Long Grove, Illinois: Waveland Press; 2016

[38] Morse JM. Mixing qualitative methods. Qualitative Health Research. 2009;**19**(11):1523-1524

[39] Miles MB, Huberman AM. Qualitative Data Analysis : An Expanded Sourcebook. London, UK: Sage; 1994

[40] Yilmaz T, Yilmazar A, Akgun S. Role of ecotourism in regional development in theory-practical axis: The case of Sapanca. ABET. 2021;**11**(Single Issue):1-18

[41] Richer I. Five best practices for future-ready tourism websites. DCI. 2016. Available from: https://aboutdci.com/2016/05/five-essential-features-future-ready-tourism-websites/ [Accessed 16 May 2023]

[42] Orîndaru A, Popescu M-F, Alexoaei AP, Căescu S-C, Florescu MS, Orzan A-O. Tourism in a post-COVID-19 era: Sustainable strategies for Industry's recovery. Sustainability. 2021;**13**:6781. DOI: 10.3390/su13126781

Chapter 5

Evolving Marketing Trends in the Era of COVID-19: A Shifted Landscape

Michael Mncedisi Willie

Abstract

The marketing arena experienced a profound shift, particularly highlighted amidst the COVID-19 pandemic, resulting in a considerable alteration of the industry's competitive dynamics. This study explores the evolving marketing trends in the era of COVID-19, highlighting the impact of the pandemic on marketing strategies and practices. It emphasizes the dynamic nature of the marketing landscape, showcasing how organizations have adapted to the challenges by embracing digital transformation, virtual experiences, and customer-centric approaches. The study also recognizes the importance of agility and resilience in navigating the changing landscape and emphasizes the need for marketers to stay abreast of emerging trends and technologies to thrive in the post-pandemic world.

Keywords: customer behavior shifts, COVID-19, digital transformation, online advertising, remote work, virtual events, social media engagement, marketing trends

1. Introduction

The COVID-19 pandemic reverberates across numerous industries, including the field of marketing, with a profound impact [1]. As the world grapples with unprecedented challenges and uncertainties, businesses are forced to rapidly adjust their strategies and approaches to align with the shifting landscape and evolving consumer behaviors [2]. This study aims to provide a succinct yet comprehensive overview of the revolutionary marketing trends that are emerging in the era of COVID-19. The study further delves into the key shifts that reshape the marketing landscape, examining how consumer behaviors and preferences undergo significant transformations during the pandemic [3]. We explore the surge in online shopping and the accelerated adoption of digital channels as traditional brick-and-mortar businesses face temporary closures and restrictions [1, 3, 4]. Additionally, This study sheds light on the heightened emphasis on trust, empathy, and purpose-driven marketing, as consumers seek reassurance and connection amidst uncertain times [5]. Furthermore, This study highlights the strategies and tactics employed by marketers to navigate these unprecedented circumstances successfully. From agile marketing practices to innovative digital campaigns, businesses embrace new methodologies to engage and resonate with their target audiences [6–9]. We examine the rise of virtual experiences, influencer collaborations, and user-generated content as effective tools for capturing consumer attention and fostering brand loyalty in a socially distanced world [6, 10]. This study endeavors to furnish marketers with a succinct overview of the dynamic marketing trends, intending to impart valuable insights and inspiration for their adaptability and success amidst the continuous challenges they face [1, 6, 11]. It serves as a roadmap to understanding the dynamic shifts and strategies that have emerged as businesses continue to navigate the ever-changing marketing landscape in the wake of the COVID-19 pandemic [1, 6, 12].

2. Background

The COVID-19 pandemic has caused significant disruptions across industries, including marketing [1]. Businesses have had to swiftly adapt to the changing landscape and consumer behavior [3, 13, 14]. The pandemic of COVID-19 has had a significant impact on the marketing strategies and practices of numerous industries [3, 6, 15]. Due to restrictions on physical interactions and increased reliance on online platforms, businesses have shifted towards digital channels, concentrating on online presence, social media marketing, and e-commerce platforms [1, 3, 4, 12, 16]. Empathetic and customer-centric marketing has become crucial, addressing the evolving needs and concerns of consumers during these challenging times [5, 6]. The crisis has also accelerated the adoption of innovative marketing techniques, such as virtual events and influencer collaborations, to stay connected with customers [6, 17]. Agility and flexibility have become vital in adjusting marketing strategies to rapidly changing circumstances [18]. The impact of COVID-19 on marketing strategies has varied across industries, with tailored approaches required for sectors such as health-care, e-commerce, and essential goods [6, 11, 16, 19].

3. Objectives

This chapter provides an overview of the evolving marketing trends during the pandemic, focusing on digital transformation, customer experience, influencer marketing, agile strategies, and virtual events and experiences [6].

The objectives of the study was to:

- Examine the influence of the COVID-19 pandemic on the evolution of marketing trends.
- Highlight the strategies employed by businesses to navigate the challenges
- Discuss the shifts in digital transformation and online presence.
- Explore the increased focus on customer experience.
- Analyze the role of influencer marketing and user-generated content.
- Discuss the importance of agile marketing and flexibility.
- Examine the rise of virtual events and experiences.

4. Study methods

This chapter provides a descriptive overview of the evolving marketing trends during the COVID-19 era [6, 20]. It is based on a comprehensive review and analysis of relevant literature, industry reports, and expert insights [20, 21]. The sources cited in this study offer insights and evidence supporting the discussed trends and strategies employed by marketers in response to the pandemic.

4.1 Impact of digital transformation and online presence

The COVID-19 pandemic has accelerated the digital transformation of businesses worldwide [1, 20]. As social distancing measures and lockdowns forced physical locations to close or limit operations, companies had to swiftly adapt their strategies to maintain business continuity [1, 20, 21]. This study compares the digital transformation and online presence of businesses before and after the COVID-19 pandemic, highlighting the key shifts and advancements that have taken place [1, 3, 4]. Before the pandemic, the evidence of advancements in digital transformation had become increasingly apparent, but its pace varied across industries [19–21]. Businesses recognized the potential of digital channels for marketing, sales, and customer engagement [3, 4, 21]. However, the adoption of digital technologies was often gradual, with many companies relying heavily on traditional brick-and-mortar establishments [6, 19, 20]. The onset of the COVID-19 pandemic served as a driving force for digital transformation, compelling businesses to swiftly adjust and enhance their online footprint [1, 4, 6, 21]. E-commerce, remote working, and virtual communication became essential for business survival [1, 16]. Companies across industries embraced digital technologies and platforms to ensure continuity, reach customers, and maintain operations [6, 21]. The pandemic significantly accelerated the growth of e-commerce [16]. With physical stores temporarily closed or operating under restrictions, businesses focused on strengthening their online sales channels [1, 3, 4]. Companies that had not previously invested heavily in e-commerce quickly pivoted, launching or enhancing their online stores to meet the surge in demand for online shopping [1, 4, 16]. Marketers shifted their strategies towards digital marketing channels as consumers spent more time online [3, 4, 22]. Digital advertising, social media marketing, and content creation gained prominence [6]. Companies invested in search engine optimization (SEO), paid advertising, and email marketing to connect with their target audience, expand brand reach, and drive online sales [4, 12]. As a result of the pandemic, remote work became the prevailing practice, resulting in a substantial upsurge in the utilization of virtual communication and collaboration tools. Companies adopted video conferencing platforms, project management software, and instant messaging applications to facilitate remote work and maintain team collaboration [6, 21]. With physical interactions limited, businesses focused on enhancing their online customer engagement [1, 4, 21]. Companies leveraged social media platforms, chatbots, and personalized email campaigns to connect with customers, provide support, and deliver a seamless digital experience [12]. Online customer service and virtual assistance became crucial elements of the customer journey [1, 4, 6, 21]. The pandemic accelerated the shift towards digital channels and online platforms [3, 4, 21]. With social distancing measures and lockdowns in place, consumers turned to the internet for their shopping needs, entertainment, and social interactions [1]. Businesses swiftly responded by ramping up their online presence, leveraging e-commerce, social media, and digital marketing strategies to reach and engage with their target audience [4, 12, 16, 22].

4.2 Influencer marketing and user-generated content

Influencer marketing and user-generated content (UGC) have become integral components of modern marketing strategies [2, 23–25]. Prior to the COVID-19 pandemic, influencer marketing was already on the rise, with brands recognizing the power of influencers to reach and engage their target audiences [23–25]. Influencers were typically prominent figures in specific niches, such as fashion, beauty, fitness, or travel. Brands collaborated with influencers to create sponsored content, reviews, endorsements, and product placements [18, 23–25]. Influencer marketing and UGC were profoundly influenced by the COVID-19 pandemic [23–25]. As people spent more time online and sought connections during periods of isolation, the importance of influencers and UGC increased [4, 23, 24]. Brands adjusted their strategies to align with the changing consumer behaviors and focused on authentic, relatable content that resonated with individuals facing new challenges [18, 23–25].

Amidst the pandemic, there was a remarkable transition towards the utilization of micro-influencers, who generally possess smaller yet actively engaged followings [18, 23–25]. Brands recognized the value of micro-influencers in building more targeted and personalized connections with consumers, as they often had niche expertise and a strong community following [18, 23–25]. Collaborating with micro-influencers allowed brands to tap into specific market segments and generate higher levels of trust [18, 23–25]. Authenticity became a key focus in the post-COVID era. Consumers sought genuine experiences and trusted recommendations from peers. UGC, including reviews, testimonials, and social media posts, gained significant traction [12, 23]. Brands encouraged customers to share their experiences and interact with their products, amplifying UGC through contests, hashtags, and community engagement [23]. This approach not only built trust but also fostered a sense of belonging and connection.

The pandemic prompted the rise of virtual influencers, computer-generated characters with their unique personalities and online presence [4]. Virtual influencers offered a controlled brand image and allowed companies to navigate physical restrictions while maintaining brand visibility [1, 21]. Brands leveraged virtual influencers to create engaging content, collaborate with real influencers, and explore innovative marketing campaigns [21]. With the rise of social media usage during the pandemic, influencer marketing and user-generated content gained significant traction [12, 23, 24]. Brands collaborated with influencers and encouraged customers to share their experiences, fostering a sense of community and authenticity [23, 24]. This approach helped generate trust, increase brand visibility, and drive engagement [23].

4.3 Agile marketing and flexibility

The rapidly changing environment necessitated agile marketing strategies [6, 7, 18]. Businesses had to be adaptable and flexible, constantly monitoring the situation and adjusting their marketing campaigns accordingly [1, 7–9]. This involved revising messaging, targeting, and promotions to align with shifting consumer needs and preferences [7–9, 18]. Agile marketing and flexibility have become crucial in navigating the dynamic and unpredictable business landscape, particularly in the context of the COVID-19 pandemic [1, 7–9].

This study explores the evolution of agile marketing strategies and the importance of flexibility in the pre and post-COVID-19 eras [7–9]. It highlights the changes and adaptations that have occurred to address the challenges brought about by the pandemic [7, 8]. Before the pandemic, agile marketing was already gaining traction as a response to the rapidly changing consumer landscape [7–9]. Agile marketing emphasized iterative and data-driven approaches, allowing marketers to adapt their strategies based on real-time insights [7–9, 21]. It involved cross-functional collaboration, continuous testing, and quick decision-making to optimize marketing campaigns [7–9]. The COVID-19 pandemic disrupted business operations and consumer behavior, requiring marketers to be even more agile and flexible [1, 18]. Companies had to quickly respond to changing market conditions, consumer needs, and government regulations. Agile marketing became essential in adjusting messaging, campaign execution, and resource allocation to ensure relevance and business continuity [1, 7–9].

In the post-COVID-19 era, businesses had to embrace flexibility and agility to rapidly adapt their marketing strategies [7–9]. This involved identifying new opportunities, revising messaging, and reallocating budgets to channels that resonated with consumers in the current context [8, 9]. Companies needed to quickly assess and respond to emerging trends and consumer sentiment. The pandemic highlighted the importance of data-driven decision making [21]. Marketers relied on real-time data and analytics to monitor changing consumer behaviors, measure campaign performance, and inform marketing strategy adjustments. Data-driven insights facilitated agile decision-making processes, enabling marketers to allocate resources effectively and optimize marketing efforts [18, 21]. The pandemic accelerated the need for companies to embrace digital channels [3, 4, 21]. Agile marketing played a critical role in quickly shifting marketing efforts from traditional channels to digital platforms [7–9]. Marketers had to adapt messaging, creative assets, and targeting strategies to engage with consumers online, leveraging the flexibility of digital platforms to reach and connect with their target audience [4].

4.4 Embracing virtual events and experiences

Physical events and gatherings faced severe restrictions, leading to a surge in virtual events and experiences [1, 21]. Marketers explored innovative ways to engage audiences through webinars, virtual conferences, live streams, and interactive online experiences [4, 6, 10]. This trend allowed brands to connect with their target market on a global scale, overcoming geographical limitations. Virtual events and experiences have gained significant prominence in the marketing landscape, particularly in the wake of the COVID-19 pandemic [6, 26]. This study explores the evolution of virtual events and experiences, examining their prevalence and impact in both the pre and post-COVID eras [6, 10]. It highlights the changes and adaptations that have occurred in leveraging virtual platforms for marketing purposes. Prior to the pandemic, virtual events and experiences were already emerging as a complement to physical events [6, 10, 21, 26]. Companies recognized the benefits of expanding their reach and engagement through digital platforms. According to Eventbrite in 2019, virtual events like webinars, live streams, and virtual conferences offered chances for people worldwide to take part, cost-effective hosting, and data-driven insights [21, 26].

The COVID-19 pandemic necessitated a significant shift towards virtual events and experiences as physical gatherings faced severe restrictions [1, 21]. Marketers quickly embraced virtual platforms to adapt their strategies. The pandemic accelerated the adoption of virtual events and experiences across industries, with businesses exploring innovative ways to engage audiences remotely [1, 26]. In the post-COVID era, marketers focused on enhancing the capabilities of virtual events and experiences. Companies invested in interactive features, virtual networking opportunities, immersive technologies, and engaging content formats [26]. They leveraged tools such as virtual reality (VR) and augmented reality (AR) to create unique and memorable experiences for attendees. Virtual events and experiences have offered unparalleled accessibility and global reach. Geographical limitations were no longer a barrier as individuals from different locations could participate without the need for travel [26]. This allowed businesses to expand their target audience, connect with customers worldwide, and foster a sense of community across borders [1].

Virtual events and experiences provided marketers with valuable data-driven insights [21, 26]. Tracking attendee engagement, behavior, and preferences became easier in virtual environments. Marketers used this data to personalize content, tailor messaging, and deliver targeted experiences. Customization based on individual preferences enhanced attendee satisfaction and improved overall event outcomes. As restrictions eased in some regions, hybrid events emerged as a popular trend [26]. These events combined the best of virtual and physical experiences, allowing for both in-person and remote participation [21, 26]. Hybrid events provided flexibility for attendees and catered to individual preferences, while still offering the benefits of global reach and data-driven insights [21, 26].

4.5 Behavioral and social norms

The COVID-19 pandemic has resulted in significant shifts in behavioral and social norms as individuals and communities adapt to new challenges [21, 27, 28]. Research shows that the pandemic has had a profound impact on various aspects of human behavior, including adherence to preventive measures, social interactions, and overall well-being [3]. Wearing masks, practicing physical distancing, and frequent handwashing have become social norms as people strive to protect themselves and others [21]. Compliance with these norms is influenced by factors such as risk perception, trust in authorities, and social influence [21]. Social interactions have also undergone significant transformations during the pandemic. Lockdowns and restrictions on gatherings have led to a shift towards virtual communication platforms and remote work setups [3]. The use of video conferencing, social media, and messaging apps has surged, facilitating remote social interactions and maintaining connections [12]. As a result, virtual interactions have become the new norm, reshaping social norms regarding communication and socialization [21]. The pandemic has also had implications for mental health and well-being, with increased levels of stress, anxiety, and loneliness being reported [21]. Consequently, there has been a gradual transformation in societal attitudes towards mental health, resulting in increased recognition and tolerance for seeking professional assistance and adopting self-care routines [21]. Conversations about mental health and the need for support systems have prompted a reevaluation of societal norms related to mental well-being [21].

It is important to note that these changes in behavioral and social norms have varied across different populations and contexts, influenced by cultural factors, socioeconomic disparities, and access to resources [21, 27, 28]. Ongoing research continues to explore the complex interplay between individual behavior, social norms, and the COVID-19 pandemic [29]. The pandemic has significantly impacted consumer behavior, including product needs, shopping patterns, purchasing behaviors, and post-purchase satisfaction levels [21, 29]. Panic buying, influenced by attitudes, subjective norms, scarcity, time pressure, and perceived competition, became a global phenomenon during the pandemic [21, 29]. Opinions and beliefs played a crucial role in shaping trends related to the pandemic, influencing behavioral intention [21, 29]. Social norms and behavioral control were identified as factors affecting attitudes and intention [21, 29]. The healthcare sector and online doctors experienced benefits from the increased online activities during the pandemic, while sectors such as recreation, travel, and hospitality were adversely affected [4, 11, 26].

5. Future trends to marketing landscape

Future trends in the marketing landscape are shaped by ongoing technological advancements, shifting consumer behaviors, and evolving market dynamics [6, 21]. These future trends present both opportunities and challenges for marketers [18]. Adapting to technological advancements, staying customer-centric, and navigating privacy regulations will be essential in driving success in the evolving marketing landscape (**Table 1**) [6].

Artificial Intelligence (AI) and Machine Learning (ML) Integration: The integration of AI and ML technologies is expected to play a significant role in marketing. These technologies can analyze vast amounts of data, personalize marketing campaigns, automate processes, and enhance customer experiences.	Voice Search and Virtual Assistants: With the rising popularity of voice-activated devices and virtual assistants, optimizing content for voice search will become crucial. Marketers will need to adapt their SEO strategies to accommodate voice-based queries and leverage opportunities for voice-activated marketing.
Hyper-Personalization and Customization: Consumers increasingly expect personalized experiences. Marketers will need to leverage data analytics, AI, and customer insights to deliver hyper-personalized content, recommendations, and offers across various touchpoints.	Influencer Marketing Evolution: Influencer marketing is expected to continue evolving, with a shift towards micro-influencers and nano-influencers who have more niche and engaged audiences. Authenticity, transparency, and long-term partnerships will be key factors in successful influencer collaborations.
Social Commerce and Shoppable Content: Social media platforms are integrating e-commerce functionalities, allowing users to make purchases directly from their feeds. Marketers will need to leverage social commerce features and create shoppable content to provide seamless shopping experiences.	Privacy and Data Protection: With increasing concerns about data privacy, marketers will face the challenge of maintaining consumer trust. Adhering to data protection regulations, adopting transparent data practices, and providing value in exchange for consumer data will be crucial.
Sustainability and Purpose-Driven Marketing: Consumers are increasingly conscious of sustainability and ethical practices. Marketers will need to integrate sustainability into their strategies, demonstrate their commitment to social and environmental causes, and communicate their purpose authentically.	Augmented Reality (AR) and Virtual Reality (VR) Experiences: AR and VR technologies present opportunities for immersive brand experiences, product demonstrations, and virtual try-ons. Marketers can leverage these technologies to engage customers and provide unique interactive experiences.
Data-Driven Decision Making: As data availability increases, marketers will need to leverage analytics tools and insights to make data-driven decisions. Real-time data analysis, predictive analytics, and marketing attribution models will empower marketers to optimize campaigns and drive better results.	Omnichannel Integration: The future of marketing lies in seamless omnichannel experiences. Marketers need to integrate offline and online touchpoints, ensuring consistency and continuity across various channels, devices, and platforms.

Table 1.
Future trends in the marketing landscape.

6. Marketing budgets and expenditure trends pre and post covid

Prior to the pandemic, global advertising spending exhibited a positive trend, experiencing steady growth on an annual basis [30, 31]. This observation is supported by several studies and reports in the field. However, the outbreak of the COVID-19 pandemic in early 2020 significantly impacted businesses worldwide, leading to uncertain economic conditions and necessitating cost-cutting measures, including reductions in advertising budgets [1].

One notable source supporting the notion of global advertising spending growth prior to the pandemic is the "Global Advertising Forecast" [32]. Published in December 2019, the report projected a positive outlook for global ad expenditure [32]. It estimated that global ad spending will increase by nearly eight in 2024, following an upward trend from previous years [32, 33]. Additionally, a study conducted by Statista, a leading market research and statistics portal, provides empirical evidence of the growth trajectory in global advertising expenditure [33].

Between 2000 and 2010, the global advertising market experienced a period of volatility characterized by significant fluctuations [33]. During this time, there were periods of substantial growth, with rates reaching as high as 11.2%. Conversely, there were also notable downturns, with lows plunging to −9.5% [33]. However, starting in 2011, the situation began to stabilize, and the growth in advertising spending maintained an average rate of around 5% [33].

In 2020, the outbreak of the coronavirus had a detrimental impact on the advertising industry, leading to a sharp decline in ad spend [33]. The decline amounted to approximately 4%, which was lower than earlier forecasts that anticipated a 9% decrease [33]. The pandemic significantly affected businesses and consumer behavior, causing advertisers to reduce their spending [33]. Despite this setback, projections indicate that the industry is expected to recover by 2024, with expenditure growth

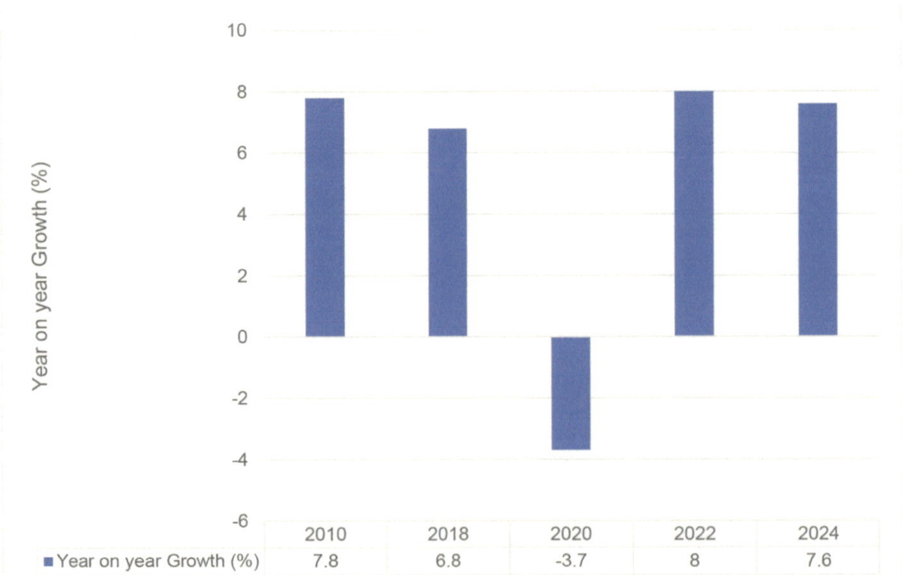

Figure 1.
Growth of advertising spending worldwide from 2010 to 2024. Source: Adapted from Statistica (2023).

returning to approximately 7.6% [33]. **Figure 1** presented a truncated trend covering the period from 2010 to 2020, along with future projections for 2024 [33] (**Table 2**).

7. Discussion

The COVID-19 pandemic has exerted a profound and lasting influence on marketing trends, necessitating rapid adaptations by businesses to accommodate the shifting landscape and evolving consumer behaviors [6, 21]. Within this comprehensive article, we will delve into the wide-ranging changes witnessed in the realm of marketing during the pandemic, emphasizing the strategic approaches employed by marketers to adeptly navigate these challenging circumstances [6]. The onset of lockdowns and restrictions on physical interactions has underscored the critical importance of digital transformation and online presence for businesses [4, 21]. The pandemic acted as a catalyst, propelling the exponential growth of e-commerce and compelling companies to augment their online sales channels [4, 16]. Recognizing the necessity of connecting with consumers who increasingly spend time online, marketers swiftly shifted their strategies towards digital marketing channels, with a particular emphasis on social media marketing and content creation [4, 12, 22]. These channels provided avenues for engagement and allowed brands to establish meaningful connections with their target audiences. Consequently, the adoption of digital technologies and platforms became not only advantageous but imperative for business survival [19, 20]. In addition to the surge in digital marketing, influencer marketing and user (UGC) gained remarkable prominence during the pandemic [22–24]. Brands promptly adjusted their strategies to align with the shifting consumer landscape and began focusing on crafting authentic and relatable content. Notably, there was a discernible shift towards micro-influencers, who possess smaller but highly engaged audiences. Leveraging micro-influencers, brands fostered targeted connections and generated higher levels of trust among consumers. Simultaneously, user-generated content in the form of reviews and social media posts gained substantial traction, fostering a sense of community and authenticity in the brand-consumer relationship [12]. Amidst the dynamic and unpredictable business landscape that emerged during the pandemic, the virtues of agile marketing and flexibility became more crucial than ever [7–9, 18]. Marketers found themselves in a continuous process of monitoring the situation and adaptively adjusting their campaigns to cater to changing consumer needs and preferences. The ability to make data-driven decisions and respond swiftly to emerging trends became paramount in ensuring business continuity [21]. Brands that prioritized these capabilities effectively navigated the challenges presented by the pandemic. As physical gatherings became restricted, virtual events and experiences emerged as viable alternatives [21]. Marketers, eager to engage audiences, explored innovative ways to connect through webinars, virtual conferences, and interactive online experiences [4, 21]. These virtual events offered numerous advantages, including global reach, enhanced accessibility, and invaluable data-driven insights [21]. Notably, the pandemic accelerated the adoption of virtual events, prompting businesses to invest in immersive technologies such as virtual reality (VR) and augmented reality (AR) to heighten attendee experiences [21]. Beyond transforming marketing practices, the COVID-19 pandemic has brought about significant shifts in behavioral and social norms [21]. Practices such as wearing masks, adhering to physical distancing, and relying on virtual communication have become deeply ingrained social norms [21]. Moreover, the pandemic has prompted a comprehensive reevaluation

Pre Covid- advertisement spending	Post Covid advertisement spend
• Prior to the pandemic, advertising spending was steadily increasing globally. • Digital advertising, including online and mobile ads, was experiencing substantial growth and had surpassed traditional advertising channels like TV and print in many markets. • Companies were investing heavily in digital advertising due to its targeting capabilities, better return on investment (ROI), and the increasing use of online platforms by consumers. • Advertisers were also allocating budgets to social media advertising, influencer marketing, and programmatic advertising.	• With the onset of the pandemic, many businesses faced financial challenges, including revenue losses, supply chain disruptions, and temporary closures. • As a result, advertising budgets were reduced or repurposed to focus on essential operations and cost-saving measures. • Industries directly impacted by lockdowns, such as travel, hospitality, and retail, experienced substantial declines in advertising spending. • Traditional advertising channels, like out-of-home (OOH) advertising and cinema advertising, were particularly affected as people stayed home and outdoor activities were restricted. • Digital advertising, especially in areas such as e-commerce, streaming services, and online entertainment, witnessed increased spending as companies sought to capture the attention of consumers spending more time online. • Programmatic advertising and social media advertising also saw shifts as advertisers adapted to changing consumer behaviors and targeted their messaging accordingly.

Table 2.
Table depicts a comparison of shift in advertisement spending: Pre and post covid.

of societal norms pertaining to mental health and well-being [21]. Individuals and communities alike have recognized the importance of mental well-being, leading to a heightened awareness and a concerted effort to address these concerns [21]. Looking ahead, the marketing landscape will undoubtedly be shaped by ongoing technological advancements, evolving consumer behaviors, and dynamic market dynamics [21]. In order to drive success in this ever-evolving environment, marketers must remain adaptable to technological changes, maintain a customer-centric approach, and skillfully navigate privacy regulations. These strategies will enable businesses to stay at the forefront of the marketing industry and thrive in the face of future challenges and uncertainties [18].

8. Conclusion

In conclusion, the COVID-19 pandemic has fundamentally transformed the marketing landscape, leading to significant changes in strategies and approaches. The crisis has accelerated the digital transformation of businesses, resulting in increased online presence and a reliance on digital channels [3, 4, 21]. Customer experience has taken center stage, with brands focusing on building stronger connections and providing personalized support. Influencer marketing and user-generated content have gained prominence, fostering authenticity and community engagement [21]. Agile marketing strategies have become crucial for adapting to rapidly changing consumer needs [7–9, 18]. Additionally, virtual events and experiences have emerged as viable alternatives to physical gatherings, offering global reach and engagement opportunities [21]. The post-COVID era presents a new marketing landscape characterized by increased digitalization, customer-centricity, and the integration of virtual experiences [21]. Marketers must continue to adapt their strategies and leverage these evolving trends to stay relevant and succeed in the dynamic business environment. By embracing digital transformation, prioritizing customer experience, harnessing the power of influencers and user-generated content, adopting agile strategies, and leveraging virtual events, businesses can navigate the challenges of the post-pandemic era and thrive in the new marketing paradigm [18, 21].

Author details

Michael Mncedisi Willie
Council for Medical Schemes, Policy Research and Monitoring, South Africa

*Address all correspondence to: m.willie@medicalschemes.co.za

References

[1] Donthu N, Gustafsson A. Effects of COVID-19 on business and research. Journal of Business Research. 2020;**117**:284-289. DOI: 10.1016/j.jbusres.2020.06.008

[2] Sharma P, Leung TY, Kingshott RPJ, Davcik NS, Cardinali S. Managing uncertainty during a global pandemic: An international business perspective. Journal of Business Research. 2020;**116**:188-192. DOI: 10.1016/j.jbusres.2020.05.026

[3] Das D, Sarkar A, Debroy A. Impact of COVID-19 on changing consumer behaviour: Lessons from an emerging economy. International Journal of Consumer Studies. 2022;**46**(3):692-715. DOI: 10.1111/ijcs.12786

[4] Szász L, Bálint C, Csíki O, Nagy BZ, Rácz BG, Csala D, et al. The impact of COVID-19 on the evolution of online retail: The pandemic as a window of opportunity. Journal of Retailing and Consumer Services. 2022, 2022;**69**:103089. DOI: 10.1016/j.jretconser.2022.103089

[5] Enslin C, Wolfswinkel M, Terblanche-Smit M. Responsible leadership through purpose-driven brand building: Guidelines for leaders in Africa. South African Journal of Business Management. 2023, 2023;**54.1**:14

[6] Pascucci F, Savelli E, Gistri G. How digital technologies reshape marketing: Evidence from a qualitative investigation. Italian Journal of Marketing. 2023;**2023**:27-58. DOI: 10.1007/s43039-023-00063-6

[7] de Weck O, Krob D, Lefei L, Lui PC, Rauzy A, Zhang X. Handling the COVID-19 crisis: Toward an agile model-based systems approach. Systems Engineering. 2020;**23**(5):656-670. DOI: 10.1002/sys.21557

[8] Lewnes A. Commentary: The future of marketing is agile. Journal of Marketing. 2021;**85**(1):64-67. DOI: 10.1177/0022242920972022

[9] Coles T, Garcia G, Turner C. Experiencing event management during the coronavirus pandemic: A public sector perspective. Frontiers in Sports and Active Living. 2022;**3**(814146):1-15. DOI: 10.3389/fspor.2021.814146

[10] Tran SK. GOOGLE: A reflection of culture, leader, and management. International Journal of Corporate Social Responsibility. 2017;**2**:10. DOI: 10.1186/s40991-017-0021-0

[11] Purcarea VL. The impact of marketing strategies in healthcare systems. Journal of Medicine and Life. 2019, 2019;**12**(2):93-96. DOI: 10.25122/jml-2019-1003

[12] Ibn-Mohammed T, Mustapha KB, Godsell J, Adamu Z, Babatunde KA, Akintade DD, et al. A critical analysis of the impacts of COVID-19 on the global economy and ecosystems and opportunities for circular economy strategies. Resources, Conservation and Recycling. 2021, 2021;**164**:105169. DOI: 10.1016/j.resconrec.2020.105169

[13] Serpil Aday S, Aday MS. Impact of COVID-19 on the food supply chain. Food Quality and Safety. 2020;**4, 2020**(4):167-180. DOI: 10.1093/fqsafe/fyaa024

[14] Redda EH. Initial impact assessment of Covid-19 on retailing: Changing consumer behaviour and

retail trade sales. JCMAN, Meyerton. 2021;**18**(2):22-41. Available from: http://www.scielo.org.za/scielo.php?script=sci_arttext&pid=S1815-74402021000200002&lng=en&nrm=iso [Accessed: May 25, 2023]. DOI: 10.35683/jcm21033.117

[15] Bieńkowska A, Koszela A, Sałamacha A, Tworek K. COVID-19 oriented HRM strategies influence on job and organizational performance through job-related attitudes. PLoS One. 2022;**17**(4):e0266364. DOI: 10.1371/journal.pone.0266364

[16] Kamel AE. The impact of COVID-19 pandemic on marketing philosophy. Delta University Scientific Journal. 2021;**4**(1):4. Available from: https://digitalcommons.aaru.edu.jo/dusj/vol4/iss1/4

[17] OECD. Digital transformation in the age of COVID-19: Building resilience and bridging divides. In: Digital Economy Outlook 2020. Supplement, OECD, Paris; 2020. Available from: www.oecd.org/digital/digital-economy-outlook-covid.pdf

[18] Hagen B, Zucchella A, Ghauri PN. From fragile to agile: Marketing as a key driver of entrepreneurial internationalization. International Marketing Review. 2019;**36**(2):260-288

[19] Gregurec I, Tomičić Furjan M, Tomičić-Pupek K. The impact of COVID-19 on sustainable business models in SMEs. Sustainability. 2021;**2021**(13):1098. DOI: 10.3390/su13031098

[20] Jaumotte and others. Digitalization during the Covid-19 crisis: Implications for productivity and labor Markets in Advanced Economies. In: Staff Discussion Note SDN2023/003. Washington DC: International Monetary Fund; 2023

[21] Moro MF, de Souza Mendonça AK, de Andrade DF. COVID-19 pandemic accelerates the perception of digital transformation on real estate websites. Quality and Quantity. 2023;**57**:2165-2181. DOI: 10.1007/s11135-022-01449-y

[22] Faruk M, Rahman M, Hasan S. How digital marketing evolved over time: A bibliometric analysis on scopus database. Heliyon. 2021;**7**(12):e08603. DOI: 10.1016/j.heliyon.2021.e08603

[23] Beveridge C. What is user-generated content? And why is it important? "User-generated content (UGC) is any content—Text, videos, images, reviews, etc.—Created by people, rather than brands". 2022. Available from: https://blog.hootsuite.com/user-generated-content-ugc/

[24] Francisco E, Fardos N, Bhatt A, Bizel G. Impact of the COVID-19 pandemic on Instagram and influencer marketing. International Journal of Marketing Studies. 2021;**13**(2):20

[25] Kim J, Kim M. Rise of social media influencers as a new marketing channel: Focusing on the roles of psychological well-being and perceived social responsibility among consumers. International Journal of Environmental Research and Public Health. 2022, 2022;**19**(4):2362. DOI: 10.3390/ijerph19042362

[26] Perez-Brumer A, Balasa R, Doshi A, Brogdon J, Doan T, Oldenburg CE. COVID-19 related shifts in social interaction, connection, and cohesion impact psychosocial health: Longitudinal qualitative findings from COVID-19 treatment trial engaged participants. International Journal of Environmental Research and Public

Health. 2022, 2022;**19**(16):10264. DOI: 10.3390/ijerph191610264

[27] Bavel JJ, Baicker K, Boggio PS, Capraro V, Cichocka A, Cikara M, et al. Using social and behavioural science to support COVID-19 pandemic response. Nature Human Behaviour. 2020;**4**(5):460-471. DOI: 10.1038/s41562-020-0884-z

[28] Singh G, Aiyub AS, Greig T, Naidu S, Sewak A, Sharma S. Exploring panic buying behavior during the COVID-19 pandemic: A developing country perspective. International Journal of Emerging Markets. 2021;**18**(7):1587-1613

[29] Hillier L. Stats Roundup: The Impact of Covid-19 on Marketing & Advertising. 2021. Available from: https://econsultancy.com/stats-roundup-coronavirus-impact-on-marketing-advertising/

[30] Holland, Knight. The Impact of COVID-19 on Your Advertising and Marketing Campaigns. 2020. Available from: https://www.hklaw.com/en/insights/publications/2020/04/the-impact-of-covid19-on-your-advertising-and-marketing-campaigns

[31] Letang V, Stillman L. Global Advertising Forecast. 2020. Available from: https://s3.amazonaws.com/media.mediapost.com/uploads/MagnaYearEnd2020Forecast.pdf

[32] Potvin Kent M, Pauzé E, Bagnato M, et al. Food and beverage advertising expenditures in Canada in 2016 and 2019 across media. BMC Public Health. 2019;**22**:1458. DOI: 10.1186/s12889-022-13823-4

[33] Statistica. Growth of Advertising Spending Worldwide from 2000 to 2024. 2023. Available from: https://www.statista.com/statistics/272443/growth-of-advertising-spending-worldwide/

Chapter 6

The Importance of Collective Brands in Protected Areas Management and Promotion: Natural.PT, Portugal Case Study

Sónia Nogueira, Catarina Mesquita, Marília Durão and Helena Albuquerque

Abstract

The objective of this study is to gain a deeper understanding of the importance of collective brands in protected areas and what is the process and impact of their development on protected areas management. This study is applied to the collective brand Natural.PT, in Portugal, using a case study approach. The study adds value to the existing literature insofar as it explores what the process of creating a collective brand involves, what benefits and/or difficulties, and what dynamics can be developed to promote this collective brand to the tourism players operating in protected areas. To this end, a qualitative study based on documentary analysis and interviews with key informants was developed, covering the different dimensions of protected areas management and collective brands. Natura.PT case is at an early stage and the main conclusions point to the need for promotion for this collective brand and a higher advantage for stakeholders allowing more decisions according to their personal and business interests and due to the fact of operating in a protected area.

Keywords: collective brands, protected areas, cooperation, tourism, destination branding

1. Introduction

The involvement of tourist destinations in brand strategy began in the 1990s, with the city of New York [1]. However, it was only at the end of that decade that the approach to destination brands reached greater expression Ritchie and Ritchie [2], fuelled by the need to compete more effectively, make decisions with greater structure, and increase accountability to stakeholders [1]. Protected areas (PA) are of particular relevance within tourism destinations, playing a crucial role in preserving biodiversity globally and conserving ecosystems while maintaining cultural values and traditional practices [3]. By establishing a recognizable identity through a collective brand, these areas can promote themselves to tourists, maintain their unique character, and facilitate collaboration among stakeholders on conservation efforts for long-term sustainability by promoting improved management and governance of natural resources [3, 4].

Concerning collective brands, the development of knowledge is in an embryonic state. The sector where there is a greater relationship with the concept is the winery [5], although there are other sectors that also need to be addressed in scientific articles. When we refer to the tourism sector, and within PA, the study of the benefits of implementing collective brands is seen as a gap, as the literature on this topic is practically non-existent.

Zhang et al. [6] argue that an increase in scientific studies on the conservation and economic development of protected areas is expected in the coming years, to improve their management. In the literature, there are many studies about the success and failure of the management of certain specific PAs [7–10]. However, regarding cooperation applied to protected areas, knowledge is still underdeveloped.

Since the tourism sector is highly relevant to the economy and due to the combination of these concepts having a high importance and impact on business management and the brand, there is a notorious gap in the literature and, for this reason, it is pertinent to choose this subject for study and deepening knowledge. Furthermore, given the current circumstances, where companies must respond quickly to consumers' desires and adapt to new realities to survive, this study will bring not only academic theoretical contributions but also practical contributions to brands that operate in PA, as well as provide support for the practical management and promotion of territories.

This research involved assembling views of experts on the collective brand Natural.PT and the Peneda-Gerês National Park (PGNP), the only protected area classified as National Park in Portugal. The research design used in this explorative research incorporated a qualitative case study approach. Methods included documentary analysis and in-depth interviews with key informants. Content analysis was used to analyze these data.

This chapter contributes to the understanding of the importance of the existence of collective brands in protected areas, as a way of leveraging the existing businesses in these regions and with that, creating a real and positive impact on territorial management. It is relevant to assess all the details of the creation process, as well as the difficulties and benefits associated with the existence of this type of brand, and the effect of their communication on regional tourism businesses, a phenomenon where still few studies have been developed [11].

2. Literature review

Regarding the definition of a collective brand, there is some agreement among authors regarding the fact that a collective brand is composed of more than one singular brand, the number that composes it can vary, although the most common is a high number of companies. The point of contention centers on decision-making. On the one hand, some authors [12] emphasize that there is total freedom in the choices inherent in business. On the other hand, some defend that the values of the collective brand must be presented [13, 14] and, accordingly, decisions must be based on common criteria regulated by a competent entity for this purpose. Therefore, a collective brand can be conceptualized as a set of unique companies that sell products and services, which may have total or partial independence in the decisions taken about their businesses and which collaborate, to extract benefits from the collective brand image.

Fishman et al. [15] mention that brands should invest in the quality of their reputation because brand recognition makes consumers formulate expectations about the quality of their products. Although the best-known brands are unique, there are many cases of collective brands where, generally, consumers are unaware of the set of singular brands that make up the same collective brand. The same authors also mention examples of products such as wine, cheese, and coffee that prove this position. These collective brands have in common the fact that they brand regional agricultural products protected by designation of origin (DOP) and geographical indication (IGP), highly regulated and certified by the competent authorities for this purpose.

The fact that collective brands are associated with superior quality results in a higher investment in quality by these collective brands, compared to individual brands. The larger the collective brand, that is, the greater the number of associated individual brands, the greater the incentive to increase the quality of products and services and, consequently, the more profitable the association of all will be, more precisely, the more profitable the collective brand to its consumers. Collective brand image affects the profit of individual companies more than their brands [12].

Collective brands, in tourism, are created to support the unique brands that constitute it, these refer to a territory or a particular tourist destination [16]. As, for the most part, tourism companies are small, associating them under a single brand can make them benefit from the transfer effect and contribute to positive synergies [17]. For the development of a collective brand, in addition to the members mutually agreeing to their association, it is also necessary that they mutually agree to share the name of the brand [12]. To achieve the success of the collective brand, all those involved must benefit from their association, making the partnership lasting. The balance between the parties is essential so that no single brand takes advantage of the other [18].

One difficulty described in the literature about collective brands is the free-riding effect. Free riding is when a given company obtains benefits from being associated with another(s), without wasting its resources. Knowing the contribution of each brand to the collective brand, it should be possible to identify inappropriate behavior and find out if there is a practice of free-riding within it, so managers must ensure compliance with minimum quality standards that do not condition or harm free-riding [16].

Collective brands, in the case of their application to protected areas, should contribute to creating sustainable development solutions, based on increasing local entrepreneurship and allowing their dissemination [19]. In addition, they can be seen as initiatives that help to achieve one of the fundamental pillars of sustainable development—governance, contributing to the increase of collaborative processes between the diverse business [20] and, helping protected areas to overcome challenges they face.

3. Methodology

To investigate the importance of the collective brand Natural.PT, a qualitative case study on Peneda-Gerês National Park—Portugal (PGNP) was developed. Data were initially collected from archival and management information sources, such as previous studies on the PGNP and open consultation documents on the Co-management Plan for the PGNP. These sources also underpinned the development of the primary tool for data collection: semi-structured face-to-face interviews.

In-depth interviews were conducted, in February 2023, with two purposefully selected members of INFC, the national organization responsible for the creation and management of the collective brand Natural.PT. Conducted in person, these interviews provided an opportunity to explore the interviewee's perceptions of the following dimensions: (C1) creation of collective brands in protected areas, (C2) dynamics of cooperation, (C3) benefits, (C4) challenges /barriers, and (C5) impact on spatial planning. The interviews were transcribed and subject to content analysis, using both a deductive approach, based on themes commonly used in the literature, and an inductive approach, considering new themes which were identified from the data.

4. Discussion

4.1 Case description: the collective brand Natural.PT and the Peneda-Gerês National Park (Portugal)

The collective brand Natural.PT is a national strategy whose implementation is the responsibility of the Portuguese Institute for Nature and Forest Conservation (INFC). This collective brand aims at the integrated promotion of the territory, the products, and the services existing in the National Network of Protected Areas and its surroundings. Its members include public and private entities that develop activities related to tourist recreational activities, accommodation, restaurants, local retail outlets, agri-food products, handcrafted products, intangible identity products, and research and dissemination projects.

The PGNP, located in the North region of Portugal, was the first protected area created in 1971 and stills being the only one that has the status of a national park. Contrary to other parks in Europe and the world, people live in the PGNP and human presence has been vital in building and maintaining the unique characteristics of this region, making this PA the most important in the country. Since its creation, and given its size and complexity, it deserves recognition for the high value of its natural and cultural heritage, combined with objectives of conservation, study, and sustainable management of its resources [21]. The PGNP follows a co-management model, which establishes strategic partners' intervention in managing these territories, such as municipalities, higher education, and non-governmental organizations that, in close cooperation with the INFC, collaborate for the sustainable preservation, development, and promotion of this protected area, its heritage, and its communities. The co-management of the PGNP is materialized through a Co-management Plan drawn up and executed by the co-management commission of this protected area. This document determines the strategy and corresponding measures and actions to be implemented to enhance and promote the territory, raising the awareness of the local populations and improving communication with all the stakeholders of this territory [22].

4.2 Creation of collective brands in protected areas

Regarding the first category (C1)—the creation of collective brands, it was divided into three subcategories, of which: (C1.1) brand objectives, (C1.2) process stages, and (C1.3) promotion.

The main objectives of creating the Natural.PT collective brand (C1.1) are to attribute value, visibility, and allow the promotion of products and services within the protected areas, and provide information about the PA to visitors, so that they, in addition to knowledge to use it, are encouraged to more conscious behaviors in these areas.

The stages of the process of creating a collective brand, in the case of Natural. PT, are: (1) conducting studies of bench branding to understand what was being done in the rest of the EU (European Union); (2) development of the brand by a single entity (INFC), intentionally, as there are already many conflicts within the PA and, with the involvement of a single entity, management becomes easier; (3) development of the brand counted with the help of other invited singular brands, representatives of certain regions, with their intervention being minimal, through meetings, so that the adherents could be made known and so that they would feel involved in the brand; (4) recognition by the Ministry and the Secretary of State and, for this reason, it obtained several financings; (5) definition of the regulation and the minimum necessary commitments (not too reductive to attract a high number of adherents); (6) allow free membership to the brand; (7) follow-up and monitoring, in a first phase, to be able to show interest and attract many people, that is, to have products and services that integrate the brand and start to work and create a network, the verification of the requirements is minimal, almost null, despite the regulation sometimes going further. Subsequently, with a high number of companies that allow joint cooperative work, mechanisms must be created to monitor and control what is or is not being done (e.g, listening through satisfaction questionnaires).

It should be noted that, in this subcategory (C1.2), it is extremely important to adapt the rules at the beginning and over time, always in line with the brand's objective, as it is possible to exclude relevant stakeholders. For example, initially, because they were associated with the State, the brand was very legalistic, asking for documents related to finance and social security debts. Later they retreated, as they realized that there is no business with the company, there is no money involved, only its promotion, that is, if the company can carry out its activity, that is enough. Another example is the fact that, initially, the regulation asked that the products be certified by another entity. Subsequently, there was a retreat in this decision, because when going to the field, it was clear that only the large companies had the financial capacity to certify their products, putting aside the adherents with local, appetizing, good quality products, but that could not pay certifications.

Something that has not yet been done by this collective brand, but which stands out as a point of improvement is the use of the co-management committee as the entity that has recognition and gives opinions to brand adherents, in other words, that monitors. The co-management committee meets monthly and, given the fact that one of its objectives is this collective brand, it would be interesting for them to verify these entities in the territory.

Concerning the promotion of the brand (C1.3), since the brand belongs to the State and the ministry is very involved and committed to its success, access to financing was facilitated. With this, the actions carried out were: (1) presence at fairs with a high number of tents and video mapping; (2) presence at events in the tourism sector with brand identification roll-ups; (3) placement of a flag with the brand's logo in the PA reception centers, to promote the council's adherent products/services; (4) sale of a plaque with the brand's logo to members, to be placed in their spaces and identification of belonging to the brand; and (5) brand guide, merchandising products, brochures with a list of all members, among other physical documents.

4.3 Dynamics of cooperation within collective brands

Concerning cooperation dynamics (C2), the brand's portal refers to "joint work" as a justification for joining Natural.PT. However, on the ground, only two activities were carried out, and these were not transversal to all geographic areas, due to lack of time and because the brand was still in its initial state. The activities carried out were:

1. In a bus, with some brand members from that region, a visit was made to the national park, with technicians who explained the dynamics in the area. With this, it was noticed that people all knew each other but, until then, they did not do business with each other. This was important to boost and initiate a network of contacts and cooperation between those companies that participated in the activity. The objective of this activity was exactly the creation of a business network, insofar as there is sharing of capabilities or even products, among others.

2. In a meeting with members of the municipality of the Guadiana natural park, they were challenged to put on a map of the park what each one valued in the PA, which could be a visiting point, a restaurant, a company that sold a certain product, or another presence in the area, if it added value. After that, they were challenged to develop a three-day itinerary, simulating being a travel agency. These scripts had a practical application, at a fair in Madrid, which had real consequences and, for this reason, the motivation for participating in this type of dynamics increased. With this action, it was noticed that there were many possibilities for creating value within the PA, which the companies that operated in the area were unaware of.

A difficulty detected for this type of dynamics is related to the fact that not all participants show the same interest in their involvement. Companies without great projection capacity are the ones that most seek ways to promote their products and services, such as tourist animation companies and accommodation. On the other hand, restaurants consider that word of mouth is enough to promote their business and that, despite being part of the collective brand, they do not value involvement in this type of initiative.

The cooperation between the singular brands that are part of Natural.PT has evidence of effectively existing, even if at a very early stage. An example of this was, by the Secretary of State, a reception of representatives from other countries, highly sensitive to the question of what is natural, where, immediately, the Natural.PT network combined efforts to gather a large sample of products. This was only possible because there was information on who the producers were and who could best represent each PA.

It should be noted that the perception of the development of these initiatives is positive and should be reinforced, as the entity responsible for managing the brand considers that, in this way, companies are open to business.

In terms of cooperation, the performance of some producer associations stands out, which has contributed to the enhancement of the territory. It would be important to create or reinforce the articulation between the sector's representative associations and educational institutions, as a way of bringing learning/training closer to the potentialities and opportunities of the territory.

Also noteworthy is the role of local development associations, which operate at various levels and with common development objectives, and which in some way have contributed to the establishment of strategic lines of action in the territory, namely about training for employability in emerging activities, support for economic activities and enhancement of local products, requalification of rural centers and built heritage, enhancement of natural and cultural heritage, qualification and tourism promotion, among others. These associations and other local, municipal, or supra-municipal institutions have been instrumental in channeling European funds and applying them to projects to boost and enhance the territory.

As a result of the existing natural and cultural conditions, tourism associated with nature and the rural environment and hydrotherapy are the most expressive segments and have been driving forces behind the development of the hotel sector, tourist entertainment activities, catering services, and small businesses.

However, seasonality and the tendency to concentrate in certain places are considered a problem, jeopardizing the sustainable and qualified development model of tourism in the region. Many of the economic and social benefits that tourism induces end up following the seasonal trend, such as the creation of jobs in the summer, which does not benefit the qualification of the workforce employed in the sector and economic and social stability.

Regarding tourist demand, there is also an inequality in the distribution of the flow of visitors in the region, with greater demand in the central area of the park. Data from recent years reveal a greater balance in the distribution of the flow of visitors, highlighting the importance of the park ports, with different dynamics of tourist attractiveness in the respective territory, but also the dynamization and dissemination of new spaces, the installation of new accommodation and tourist infrastructure outside the main centers of demand.

In general, the existing accommodation units in the region are of high quality. However, there are very few accommodations recognized as nature tourism ventures and Natural.PT. For the promotion of accommodation, there are currently new platforms, more current and in line with the needs of tourists, which work with the dissemination and sale of accommodation and tourist programs throughout the park.

Concerning tourist entertainment activities, in particular nature tourism, the PGNP presents a very significant offer. However, the collective brand Natural.PT still has little expression. The PA co-management model is seen as an opportunity to strengthen partnerships and establish common commitments in terms of the promotion and sustainable enhancement of the territory, heritage, and communities.

The qualitative information gathered allowed us to understand that the actual PGNP Co-Management Plan incentives associations and cooperative dynamics that contribute to the development of the territory, try to generate identity value of the territory for the local population, which has contributed to the appreciation of the patrimony; considers initiatives for employability in emerging and strategic activities for the development of the territory and progressive and qualitative growth of the tourism sector in the park, especially in terms of the offer of tourist accommodation and nature tourism activities.

However, the territory does not have yet enough resources for effective management and surveillance of the different uses of the territory, an absence of a system for surveying and recording intangible heritage throughout the territory, and a global centralized project for the interpretation of relevant cultural heritage elements. The qualitative study highlighted: (i) some failures in the promotion, communication, and positioning of the PGNP referred during the interview and overlapping actions of the various local entities; (ii) lack of safety information regarding the risks inherent to visitation; (iii) lack of a specific framework for tourist entertainment companies working in the PGNP; (iv) inadequacy of the tourist offer for different target types;

(v) difficulty in monitoring and managing visitors; and (vi) a lack of information and interpretation strategy on the natural and cultural heritage.

The co-management strategy is based on a participatory model, which is intended to create a proximity and collaborative management dynamic, involving all those who are part of or who can contribute to the sustainability and enhancement of the territory. In this way, the involvement and participation of stakeholders of interest in the constructive discussion of proposals and concrete actions constitute a fundamental process in the management of PAs.

Being an exercise in strategic planning, consultation with the various stakeholders was carried out at different times, namely:

i) public sessions to present the PA co-management model to the municipalities;

ii) online survey on the ADERE (Association for the Development of PGNP) website;

iii) participatory sessions with key actors in the territory;

iv) public consultation;

After collecting all the contributions, it is analyzed and the management plan will be prepared. Following the objectives defined in the Co-management Plan, three strategic axes were established, which accommodate initiatives that meet the promotion of PGNP, socioeconomic and sustainable development use of the PGNP [21].

4.4 Benefits, challenges/barriers of collective brands

From the perspective of the entity that manages Natural.PT, the enormous benefit (C3) is fundamentally the possibility of disclosing participating companies on the platforms of the collective brand. Only at the level of perception, the adherents felt that Natural.PT was advantageous for them since it originated, many people were interested, and it is believed that they were encouraged by seeing the involvement of those who were already part of the collective brand. Proof of this appreciation of business by the brand is the fact that adherents often question the responsible entity about the management of the brand.

Another benefit is the fact that a connection was created with the entity responsible for managing the PAs, far beyond the collective brand, what is now a facility to deal with conflicts that may exist within the scope of business. Cooperation goes beyond the brand; it has to do with the training provided and the information exchanged with members. In other words, adherents are oriented toward whom to approach and how to approach them to manage their businesses.

The degree of satisfaction of stakeholders with the brand has never been heard before. It is believed to be premature, given the state in which its development is found, lacking growth. However, it is considered relevant, as it is not yet possible to recognize whether people looking for participating companies do so through prior contact with Natural.PT. This would be interesting to measure, as it would be a good way to understand the effect of the collective brand on visitation.

On the part of the stakeholders for publicizing the brand, thet put the Natural.PT brand logo on the communications used. The companies themselves aroused interest in others and regional entities and the brand spread.

Specifically in Guadiana, instead of creating a new brand, the brand's image was used, and a new project called Natural.PT Guadiana was founded, and there was a large investment in promotion, people gathered for workshops, among other initiatives.

The great importance of the involvement of the municipalities in the question of the brand, when holding events and fairs, is highlighted. If there is an invitation for Natural.PT to participate, the participating brands may be present and, consequently, promote the awareness of their products and services.

Regarding the visibility of Natural.PT, it is considered that everything related to nature is more sought after and sold, that is, there is a great propensity to associate the word "natural" with products and services. It should be noted that, in the Azores, there is no natural park. However, despite the PA classifications not being the same as on the mainland and given the importance attributed to what is natural, as a way of adding value to PA communication, the set of PA on each of the archipelago's islands was given the name of Natural Park. This is from the point of view of communication, of selling an image, this terminology is the strongest abroad.

For those responsible for the brand, having it visible that the origin of the product comes from a PA is more important than placing the collective brand that covers it, as they claim that the purpose of the brand is to identify products with the PA. These indications for members may prove to be a weak point, compromising the achievement of high brand awareness. In addition, it is believed that there is still a long way to go in terms of Natural.PT's notoriety because the brand and its members cannot demonstrate how advantageous this association is, and its interest in attracting other stakeholders for it.

On the other hand, the challenges/barriers (C4) are: (a) the lack of doors, in most regions, which prevent the monitoring and counting of visitors, which makes the action in the PA difficult; (b) despite adherence, the lack of contact points with the brand sometimes causes a feeling of distance with the PA, as the structure of the responsible entity does not have anyone physically allocated to the regions; (c) people are afraid to put the word "natural park" on their products, due to the authorization they may require from the park; (d) Natural.PT suggests placing information relating to PA on companies' online platforms, the problem is that some companies are so small that they have the management of their online platforms contracted to external entities and when any change is necessary; however, small it may be, these companies have to pay for the changes and then back off; (e) at this time, the collective brand platform accepts new members but, after a certain period, eliminates them; (f) there is no verification for entities with fewer environmentalist practices associated with the protection, there is the only incentive for more correct behavior, rectification, going toward more sustainable practices; (g) there are no legal measures that prevent the action of brands in the PA, there is a regulation to be complied with, but the control is only on the minimum requirements. If they are not reached, those responsible get in touch with those who propose to join and communicate what should be changed, improved, or added, after a certain time, eliminate them.

4.5 Impact of collective brands on territorial management and planning

According to Environmental Fund [22], in a sustainable base model, it is necessary to adapt and innovate so that the territory is better prepared for new challenges and can thus create more and new opportunities for development, valuing people and their legacy and promote more proactive management of nature conservation, biological and geological diversity, and the landscape. It is in this context that co-management fits, a model of proximity management, which enshrines the intervention of strategic partners in the management of this territory which, in close articulation with the INFC (Institute for Nature and Forests Conservation), assumes the commitment to collaborate for the promotion and sustainable enhancement of the territory, its heritage, and its communities. The implementation of the PA co-management model implies the elaboration of a Co-management Plan.

According to Melgaço [23], the proposed PGNP Co-Management Plan, from 2022 to 2027, has the following objectives:

a. The creation of a shared dynamic for valuing the PA, based on its sustainability and focusing specifically on the fields of promotion, awareness, and communication.

b. Establish concerted procedures, aimed at better performance in safeguarding natural values and in responding to society's requests, through greater articulation and efficiency of interactions between the INFC, municipalities, and other competent public entities.

c. Generate a closer relationship with citizens and relevant entities for the promotion of the sustainable development of the PA.

The public and associative sector has sought to publicize and increase the attractiveness of the cultural heritage, with a view to enhancing the touristic value of the territory. Even so, there is a lack of complementary heritage interpretation initiatives that facilitate the perception of its value and its enjoyment by the local population and visitors. In terms of promotion, a mainly local strategy is maintained, with campaigns and publications driven mainly by the municipalities, to the detriment of a joint communication that, perhaps, could be more efficient in terms of tourist promotion and have more impact and return for the territory. The collective brand Natural.PT could give extra help with this objective. The perception of the quality and economic value of local products/services is an objective that can also be applied to traditional knowledge and practices incorporated into products and human activities carried out, in a strategy to enhance existing resources. In addition, it can serve as a strategy to leverage and differentiate other expanding areas such as nature tourism, restoration, and handicrafts, among others.

Finally, to understand the impact on territorial planning (C5), the collective brand Natural.PT is present in several structures, which does not allow its extinction because it is part of INFC's PA co-management division, integrating the co-management commission, as one of its indicators (there are objectives concerning the number of Natural.PT adherents per year).

The great advantage of the co-management committee relates to the fact that communication between all those involved allows for complementarity and avoids wasting resources. Previously, there could have been multiple brands associated with the PA. At this moment, everyone competes to strengthen the Natural.PT collective brand.

5. Conclusion

Probably implied by the fact that Natural.PT is still at an early stage, there is a need to create new dynamics for this brand. Although each stakeholder makes decisions according to their personal and business interests, the information actively disclosed by the collective brand means that people cannot claim that they were not informed about how to act more responsibly within the PA. Because joining the brand has no cost for members, it is believed that the country's economic situation has no impact on Natural.PT.

The primary point will be to attract new members, an objective already established in the co-management plan. Understanding the stakeholders involved in the collective brand will also be a relevant topic in brand management and, consequently, in its success, as listening to those on the ground helps to understand the real points of improvement in the business and the difficulties faced in the PA.

At this moment, it remains to continue the brand's cooperative actions, although two activities have already been carried out, it is necessary that they cover all protected areas, transversally, and that they have more consistency. Only then will it be possible for PA members to get to know each other, gain confidence, and start networking (this is one of the main objectives of the brand).

One of the main points to leverage this brand is the investment in promotion, as this is the only way to boost visitation. Despite the existence of Natural.PT, it is not possible, until then, to monitor whether people looking for members arrive because they have been in contact with the collective brand. A suggestion for this practice will be the use of digital branding strategies, such as investing in social networks, and email branding, among others.

The co-management plan is evaluated according to a minimum set of indicators (21, in this case), defined in an ordinance. Each PA can, to these indicators, and add others, but these must be included. The mission of the indicators is to assess how executable the plan is, in other words, how well the objectives defined in it are being achieved. Normally, the indicators have a starting point, which is the moment when the evaluation begins, and an endpoint, which defines the effectiveness and efficiency of the action. The minimums are established by each PA and are assessed annually.

One of these indicators is exactly the number of adherents to the collective brand, which means that, in the plan, there must be one or more actions that make the connection to the collective brand attractive and desirable.

Another obligation of the co-management division is the promotion of the PA. The PA is greatly promoted when it has the scope of the collective brand with a certain quality, with singular brands that assert themselves in a certain way. With this, a relevant topic is the search and identification of stakeholders. Those already involved in the collective brand were companies that are concerned with this issue and are committed to it. However, this topic should maintain consistency over time and continue to evolve.

Acknowledgements

This work was supported by the UIDB/05105/2020 Program Contract, funded by national funds through the FCT I.P.

Conflict of interest

The authors declare no conflict of interest.

Author details

Sónia Nogueira[1*], Catarina Mesquita[2], Marília Durão[1,3] and Helena Albuquerque[1,3]

1 REMIT (Research on Economics, Management and Information Technologies), Portucalense University, Porto, Portugal

2 Minho University, Portugal

3 GOVCOPP (Research Unit on Governance, Competitiveness and Public Policies), Aveiro University, Aveiro, Portugal

*Address all correspondence to: snogueira@upt.pt

References

[1] Morgan N, Pritchard A, Pride R. Tourism places, brands, and reputation management. In: Morgan N, Pritchard A, Pride R, editors. Destination Brands. 3rd ed. Portsmouth: Butterworth-Heinemann; 2011. pp. 3-19. DOI: 10.1016/B978-0-08-096930-5.10001-1

[2] Ritchie JRB. The branding of tourism destinations – Past achievements and future challenges. In: Annual Congress of the International Association of Scientific Experts in Tourism. Marrakech, Morocco; 1998

[3] Dudley N, editor. Guidelines for Applying Protected Area Management Categories. Lisbon: IUCN; 2008

[4] Chidakel A, Eb C, Child B. The comparative financial and economic performance of protected areas in the Greater Kruger National Park, South Africa: Functional diversity and resilience in the socio-economics of a landscape-scale reserve network. Journal of Sustainable Tourism. 2020;**28**(8):1100-1119. DOI: 10.1080/09669582.2020.1723602

[5] Castro VA, Lourenção MT, de Giraldi J, JHC O. Creation and implementation of collective brands: An analysis of the Brazilian wine sector challenges. Journal of International Food and Agribusiness Brandeting. 2021;**35**(1):1-19. DOI: 10.1080/08974438.2021.1924334

[6] Zhang Y, Xu J, Yao Y, Yan Z, Teng M, Wang P. What is the relationship between natural protected areas and stakeholders? Based on literature analysis from 2000-2021. Forests. 2022;**13**(5):1-20. DOI: 10.3390/f13050734

[7] Orange JM. Nature Tourism Management in the National Network of Protected Areas: The Nature Sports Charter of the Sintra-Cascais Natural Park. Lisbon: Universidade Nova de Lisboa; 2011

[8] Miles A, Perea Muñoz JM, Bayle-Sempere JT. Low satisfaction and failed relational coordination among relevant stakeholders in Spanish Mediterranean marine protected areas. Journal of Environmental Management. 2020;**272**:111003. DOI: 10.1016/j.jenvman.2020.111003

[9] Syrou D, Botetzagias I. Stakeholders' perceptions concerning Greek protected areas governance. Sustainability. 2022;**14**(06):3389. DOI: 10.3390/su14063389

[10] Tumusiime DM, Vedeld P. False promise or false premise? Using tourism revenue sharing to promote conservation and poverty reduction in Uganda. Conservation and Society. 2012;**10**(1):15-28. DOI: 10.4103/0972-4923.92189

[11] Mas-Ruiz FM, Sancho-Esper F, Sellers-Rubio R. The effect of collective brand on advertising productivity. British Food Journal. 2016;**118**(10):2475-2490. DOI: 10.1108/BFJ-01-2016-0032

[12] Fishman A, Finkelstein I, Simhon A, Yacouel N. The economics of collective brands. SSRN Electronic Journal. 2010:1-30. DOI: 10.2139/ssrn.1317262

[13] Fernández-Barcala M, González-Díaz M. Brand equity in the European fruit and vegetable sector: A transaction cost approach. International Journal Of Research in Marketing. 2006;**23**:31-44. DOI: 10.1016/j.ijresmar.2006.01.004

[14] Rangnekar D. The Socio-Economics of Geographical Indications – A Review

of Empirical Evidence from Europe [Internet]. In: Project on Intellectual Property Rights and Sustainable Development. Issue Paper No. 8. Geneva: ICTSD (International Centre for Trade and Sustainable Development and UNCTAD (United Nations Conference on Trade and Development). Available from: https://unctad.org/system/files/official-document/ictsd2004ipd8_en.pdf

[15] Fishman A, Finkelstein I, Simhon A, Yacouel N. Collective brands. International Journal of Industrial Organization. 2018;**59**:316-339. DOI: 10.1016/j.ijindorg.2018.03.002

[16] Nicholas JL, Mas FJ. Detecting free riders in collective brands through a hierarchical choice process. Journal of Travel Research. 2015;**54**(3):288-301. DOI: 10.1177/0047287513517419

[17] Aarstad J, Ness H, Haugland SA. Network position and tourism firms' co-branding practice. Journal of Business Research. 2015;**68**(8):1667-1677. DOI: 10.1016/j.jbusres.2015.03.025

[18] Carvalho C. Development of International Collective Brands in the Footwear Sector [Master's Dissertation]. Ponte da Barca: Institutional Repository of the University of Minho; 2014

[19] Matos Silva F, Sousa C, Albuquerque H. Analytical model for the development strategy of a low-density territory: The Montesinho Natural Park. Sustainability. 2022;**14**(7):4373. DOI: 10.3390/su14074373

[20] Ferreiro MF, Sousa C. Governance, institutions and innovation in rural territories: The case of Coruche innovation network. Regional Science Policy and Practice. 2019;**11**:235-250. DOI: 10.1111/rsp3.12147

[21] Environmental Fund. PGNP Co-Management Plan Proposal – Characterization of the Protected Area. 2022

[22] Adere PG. Co-management of the Peneda-Gerês National Park. Adere Peneda-Gerês. Lisbon: The Interne. 2020. Available from: https://www.adere-pg.pt/pt/cogestao

[23] de Melgaço M. Co-management plan for the Peneda Gerês National Park. Ponte da Barca: The Interne; 2022. Available from: https://www.cm-melgaco.pt/plano-de-cogestao-do-parque-nacional-da-peneda-geres-consulta-publica/

Chapter 7

Tourism Cohabiting with a Pandemic: Lessons from the COVID-19 (2020–2023)

Alain A. Grenier

Abstract

The pandemic caused by the striking transmission of COVID-19 in early 2020 decimated the population by attacking the most vulnerable—those with chronic health problems and the elderly: one of tourism's most important clientele. Before the authorities had the tools to treat and protect the population from this virus, unprecedented sanitary measures were imposed in most countries of the world, restricting freedom of movement—the core of the tourism experience. If no economic sector was spared, tourism was among the hardest hit. As a luxury product, tourism was the first sector to suffer the repercussions of political, economic, environmental and health crises. Economic downturns usually result in layoffs and loss of revenues. The COVID-19 crisis also led to the partial "destructuring" of the tourism industries. In those circumstances, the challenge was to maintain just enough tourism activity to save the enterprises and services involved while preventing the spread of the disease any further. The crisis exposed the fragility of the tourism industry's capability to adapt and cope with a sanitary crisis. Based on experiences identified in the literature during the pandemic, this study proposes an overview of the adaptation strategies deployed by the tourism industries. The study aims to pinpoint resilient avenues for dealing with future health crises.

Keywords: tourism, sanitary crisis, pandemic, lessons, management

1. Introduction

In terms of its nature—the transmission of a virus, the crisis generated by the COVID-19 pandemic in 2020–2023 was nothing new. It was the lightning transmission of the disease, its severity—fatal for many, the absence of treatment for the first year of the pandemic, and the shutdown of the world economy—including tourism—that left their mark on the collective imagination. As an agent of virus transmission, tourism was no innocent player in the disaster. It was not the first time. In the Middle Ages, it took travellers 10 years to spread the bubonic plague (Black Death) to the four corners of Europe and North Africa. The great influenza ("Spanish") pandemic of 1917–1918 took a year to reach Europe from the USA, carried by soldiers in the midst of war. In 2020, the COVID-19 virus took just over a month to spread from China—

where it is believed to have originated—to the entire planet. The enlightened transmission of viruses cannot be dissociated from technological advances in transport and human mobility, as evidenced by the effects produced by a billion cross-border journeys across the planet every year [1, 2]. Travel (business) and tourism (leisure travel) have thus become both the main agents of propagation and its main victim. By April 28, 2020, 2 months after its emergence on American soil, COVID-19 had claimed as many victims (58,365) in that country as the Vietnam War had in its 20 years of conflict (1955–1975) [3].

Before the pandemic, travel and tourism (including its direct, indirect and induced effects) were considered to be one of the world's largest sectors, "accounting for 1 in 4 of all new jobs created worldwide", and credited with "10.3% of all jobs (333 million) and 10.3% of global GDP ($9.6 trillion)" [4]. In short, in many places, overcapacity was leading to situations of over-tourism—the saturation of tourist sites and residences when the number of tourists exceeds one of the dimensions used to measure carrying capacity (physical, environmental, visual, noise and/or socio-cultural), leading to the deterioration (or even exhaustion) of resources, the quality of the expected recreational tourism experience and/or the quality of life of the host populations. Despite pre-pandemic appearances, tourism is a fragile business sector that is sensitive to the vagaries of economic, political, environmental and health conditions, both locally and internationally.

When the World Health Organisation declared the COVID-19 pandemic on March 11, 2020 [5]—for a chronology of events, see [6–8], the governments of the richest countries decreed the shutdown of economic sectors deemed non-essential and imposed an initial two-week confinement on their populations. No one could have imagined at the time that evil would strike for 2 years and that the crisis would extend well beyond this period, generating inflation and even recession. In the space of 2 months, the globalised economy stopped spinning. Among the collateral victims were the tourism industries, which were also condemned to confinement. While it may seem obscene, at first glance, to bemoan tourism when healthcare workers are giving their all to care for the sick in failing conditions—many at the cost of their lives at a time when there is little understanding of how the virus spreads and no cure yet exists—other human dramas are playing out behind the closed doors of containment. From 1 day to the next, entrepreneurs, tour operators, hoteliers, innkeepers, restaurateurs and suppliers, attraction managers, their employees and their families are faced with uncertainty: there are no more tourists, no more customers and no more income. It took only 2 months after the outbreak of the SARS-CoV-2 (COVID-19) virus for the tourism industry to collapse as borders were closed, travel and even socialisation banned—unheard of in peacetime.

The pandemic put tourism on hold for an indefinite period. A luxury product, tourism requires mobility and socialisation—two pillars that are proving to be the main vectors for cross-border transmission of the disease. State-imposed restrictions on both mobility and socialisation generated anxiety among the population, and even domestic violence. This is a reminder of the importance of leisure, to which tourism belongs. As a means of asserting social and even identity, the tourist experience provides customers with physical and psychological benefits through contact with and discovery of new environments. As a phenomenon of post-industrial societies, tourism can, when the time comes, be seen as a solution to the anxiety problems of populations, rather than as an obstacle, at least if it is practised domestically. This will suit certain destinations with a local clientele. However, in emerging countries such as the Caribbean, where the clientele is essentially made up of foreign tourists, the barriers to mobility are even more catastrophic. As a result, in 2020 alone, the pandemic caused the loss of 62 million jobs, which represented nearly US$4.9 trillion and a 50.5% reduction in tourism contribution to GDP ([4], p. 8).

The near-complete cessation of tourism, particularly international tourism, raised the spectre of destructuralisation, i.e., the permanent reallocation of key industry infrastructures—such as transport, accommodation and catering—to functions other than those for which they were designed (residential housing, for example). When tourism resumed, it was feared that some infrastructures (transport and accommodation, mostly) would no longer be available and that the industry would shrink. The dependence of some economies on tourism alone also showed its limits. Foreign tourism had to be part of the solution, along with economic diversification.

Not only does the pandemic demonstrate the fragility of national and global economies, but it also sheds light on shortcomings in the degree of preparedness of states and the ability of their societies to cope with a crisis of this scale. Some researchers [9–11] have addressed health crises in their work, particularly in relation to the SARS pandemic (for instance, [12]). However, none of their studies covered the management of such a universal crisis as the one caused by COVID-129. Some studies have looked at the effects of epidemics, i.e., localised to a given area. Studies on the impact of international mobility have tended to focus on the (involuntary) import of foreign plant and animal species (see [13]). The effects of a pandemic on tourism will not be studied until 2020. Hence, we found ourselves "building the plane in flight" to use the image coined in 2016 by a Google director, Max Ventilla.

The crisis also raised doubts about societies' ability to respond to other crises of the same magnitude in the possible event of multiple crises—combinations of sources of disruption [14]. Yet recent history has left clues in its wake as to the possibilities of global crises: the terrorism attacks of September 11 (2001), the avian flu (SARS, 2002–2004), the H1N1 virus (2009), the Middle East Respiratory Syndrome (2012) and the Ebola haemorrhagic fever in Central and West Africa (2013–2016). Apart from the political terrorist attacks of September 11, the other major events were all health-related. They all reached the epidemiological level (contagion of a large number of people in a given region where the disease does not usually prevail) or the pandemic level (contagion that spreads to the entire population of several countries or even beyond a continent) without ever reaching the planetary threshold.

The impacts of the COVID-19 pandemic are widely described in both grey and medical literature. Tourism intellectuals were fast not to be outdone, launching as early as March 2020 special issues written on the spur of the moment, almost unanimously calling for a rethink of the way the tourism industries operate, arguing for more equitable and sustainable consumer choices (see [15, 16]). Some even called for the "deglobalisation" of tourism (see [17]). Wishful thinking if ever there was, but disconnected from the economic reality facing tourism entrepreneurs, many of whom have been bled dry by this pandemic that they never saw coming and against which few, if any, had been able to immunise themselves financially. This raises the question of the present: how to survive, while waiting for the aftermath—and how to get there? There is also the question of tomorrow: how can the tourism industry recover from a global crisis, knowing that transport, the essential pillar of travel, is also its Achilles heel as a vector for cross-border transmission of the disease?

Drawing on the experiences identified in the literature during the pandemic (notably [18]), this study provides an overview of the adaptation strategies deployed by tourism industries around the world to maintain, at the very least, leisure travel during the COVID-19 pandemic. It aims to identify resilient avenues in the face of future health crises.

To this end, the study:

a. juxtaposes the conditions necessary for tourism with the health challenges of a pandemic;

b. identifies the conditions required for business recovery;

c. lists responses from the tourism industries.

2. Methodology and study limits

Aware that each tourism sector requires specific solutions, the author has chosen an inductive approach—reasoning that moves from the specific to the general [19]. Thus, "from reported or observed facts [...], the researcher arrives at an idea by generalisation rather than by verification against a pre-established theoretical framework", underline Blais and Martineau ([19], p. 5). The study draws its data from the story of the crisis and the measures put in place by the different types of actors (governments, entrepreneurs, support service providers, etc.), in order to draw generalisations applicable to the whole, as far as possible. At this stage of the crisis, the present study has opted for findings based on grey literature, i.e. press articles, which abound on tourism issues, illuminated by previously published studies on the conditions necessary for tourism development.

A single study cannot claim to be able to bring all these lessons together—each player and each environment has its own context and particularities. Beyond these micro considerations, however, there are macro principles that can provide clues as to the scenarios to be envisaged in the event of the next global health crisis. The article summarises the types of adaptations employed by the tourism industry in a context of viable cohabitation with the evil that requires it to review the shape of its products and services—in short, its modes of operation, while recognising the volatility of scenarios in the event of a new virus with unprecedented effects.

3. Conditions for tourism

A remedy for the alienation of industrialised and post-industrial societies, tourism involves moving away from one's domestic and familiar world to a foreign, national or international host environment. This displacement from one's mastered universe leads to a loss of reference points, which, in the context of the support provided by the host, encourages the renewal of mind and body—the vacation. Six conditions must be met in order to carry out a tourist activity:

1. the existence of an attraction (which stimulates the desire, the dream, the project and finally the decision to go);

2. a discretionary budget (what is left in the budget after you have taken care of your basic needs: housing, food, etc.: to cover the financial costs of your dreams): to cover the financial costs of your dreams)

3. an infrastructure to support the dream (transportation, lodging, field guides, etc.);

4. paid leave (time off from regular and funded activities)

5. health (being able to get around and enjoy your stay away from home); and

6. safety (guaranteeing healthy physical and psychological conditions for the trip) and its perception [20]

When these conditions are met, the majority of people can indulge in leisure travel (tourism). Tourism allows people to renew themselves through immersion in new environments and the realisation of new experiences involving varying degrees of socialisation. This socialisation inevitably requires physical proximity (proximity of service) and interpersonal exchanges, ideal for the transmission of viruses that can lead to disease. However, by compromising four of the conditions necessary for travel (health and safety, loss of discretionary budget and paid leave, for many), the 2020–2023 COVID-19 pandemic has forced a halt to all activities requiring physical proximity. Tourism could only resume under certain conditions (6), as set out by the World Health Organisation (**Box 1**) [21].

Criteria 2, 4, 5 and 6 were particularly relevant to tourism. The second involved guaranteeing the presence of an unsaturated healthcare system capable of treating all patients. The precautionary principle thus requires each tourist to be a responsible citizen, so as not to create or contribute to situations that could exert pressure on the health system of their territory, or that of a receptive environment, by risk of importing the disease (criterion 5). Criteria 4 addressed the issue of space organisation (including ventilation, as we would learn) to prevent proximity contact and the transmission of the virus. Criteria 6 stressed the necessity of population awareness of the risks and the precautionary behaviours to be adopted.

Those steps were required as tourism is a giant with feet of clay: it is sensitive to the vagaries of economic, social, political and health moods. In the absence of a pharmaceutical solution that could be administered to vast swathes of the human population, tourism became virtually impossible across borders. With the arrival of a pharmaceutical solution (1 year into the COVID-19 pandemic, at the end of 2021), cohabitation with the virus replaced the wish for community immunity, which would

ONE: transmission must be controlled.
TWO: capacity of the healthcare system to detect, test, isolate and treat each case and trace contacts.
THREE: minimising the risk of epidemics in specific locations, such as healthcare facilities and nursing homes.
FOUR: implementation of preventive measures in workplaces, schools and other places essential to the population.
FIVE: control of import risks.
SIX: raising awareness and involving the population to equip them with the means to adapt to the "new normal".

Source: adapted from WHO [21].

Box 1.
Six criteria for lifting containment restrictions of the 2020–2023 pandemic.

not come [22]. Community immunity requires a sufficiently high number of immunised individuals to limit contagion of the disease to less than one person per agent of transmission (i.e., an R0—reproduction zero—of less than 1). This rate varies according to the degree of transmissibility of the disease. Rawaf et al. [23] predicted nine (9) waves of transmission (as of fall 2022, 8 waves of COVID-19 had been recorded in many countries [24, 25]). Economic recovery therefore took place against a backdrop of partial containment including tourism market bubbles, i.e. by agreement between specific populations where contagion was under control, if not limited, creating a gigantic diplomatic and transportation headache. The challenges facing the tourism industry were therefore manifold: restoring customer safety and confidence by adapting services and infrastructures to the new health regulations, while ensuring the sustainability of operations, without having to rely on the profit margins enjoyed by some companies up to 2020. The answers would come from the resilience of some and the imagination of others.

4. Crisis management

Sooner or later, every unit, a group, a society, a company or an organisation is bound to experience a crisis. A crisis can be defined as "a process in which an unusual, unexpected, or the least-expected situation occurs in a certain period." ([26], pp. 2–3). Moreover, this situation temporarily disrupts the proper functioning of the unit, requiring an active response on its part. There are several levels of crisis, from minor to major, including catastrophe—"[an] unexpected discontinuity in an otherwise continuous system," ([27], p. 255) which leads to irreparable loss. To survive, the unit must reinvent itself through fundamental change. The crisis evokes the setback or possible setback for the unit due to major, often unexpected elements against which it is unprepared or ill-prepared. It requires the unit to react to choices of a behavioural, ethical or moral nature: resilience. This response may even call into question the organisation of the unit itself. As Ritchie et al. ([28]: p. 203) point out, understanding the nature of a crisis is important. However, it is even more important to know how to manage it in order to mitigate its impact on the organisation and its short—and long-term development. A unit unable to manage a crisis is bound to disappear and be replaced by another. "[T]he real challenge is to recognise crises in a timely fashion and implement coping strategies to limit their damage," argues Darling et al. [29].

Two elements are particularly important in this management: communication (the message and how it is delivered) and perception (how events and the message are received and understood by the public affected by the crisis). In tourism, communication helps destinations both to limit the negative impacts of a crisis and, at the same time, to recover from the difficulties that have arisen ([28], p. 203). However, there is a third element: the revival of the service damaged by the crisis.

Crisis scenario planning can help prevent a system or organisation from imploding. It requires cold thinking and consideration of the responses to be given to different situations.[1] These situations can be considered individually or in combination, i.e., a multidimensional crisis. These include crises of an intra-organisational nature—within an organisation—and extra-organisational crises—external to the organisation,

[1] See Grenier ([30]: pp. 399–403) for a list of risks in tourism.

but whose effects have a direct impact on it. In the case of the 2020–2023 health crisis, it began extra-organisational and rapidly became intra-organisational.

Responses to crises and disasters, although conceivable, can never be fully planned. Actors can prepare themselves to prevent the implosion of their system or organisation. This means considering and thinking, in the cold light of day, about responses to different scenarios. These include crises (internal to the organisation) and catastrophes (external to the organisation). One can respond to a disaster insofar as the players still have the resources to do so, i.e., the supply networks (services and raw materials). If these are destroyed, or if access to them is cut off, disaster management inevitably requires help from outside the affected area. The organisation must therefore develop mutual aid networks in anticipation of disasters while managing its own (implicitly internal) crises. A crisis takes on unprecedented proportions when it affects both the actor and its networks. In a way, this was the COVID-19 pandemic scenario.

Any crisis calls into question the notion of safety—the means available to a player to deal with risks, in addition to those resulting from bad intentions (in which case it is a question of security). The same applies to tourism. However, the usual security measures deployed in the tourism industry proved useless against the virus, as its players had mainly invested their efforts, since September 2001, in security measures targeting criminal activity—crimes against people and property. In terms of health safety, the tourism industry's actions were often limited to that of premises and equipment in the face of environmental and behavioural risks. With the exception of food hygiene and vaccination requirements for some destinations, nothing really addressed health safety.

For tourists, safety is a question of both perception—whether or not they are in danger (target) or exposed to danger (collateral victim)—and evaluation of the actual risk. For example, states afflicted by social or even armed conflict may continue to receive tourists if the crisis does not affect the area where holidaymakers are concentrated. In the case of a health crisis, few places can claim not to be at risk, since the tourist—the beneficiary—is himself the potential vehicle for disease transmission. Kanlayanasukho and Veuthey ([31], p. 108) propose two types of crisis management response:

a. the operational response—the taking of immediate decisions and actions to manage the effects of a given crisis on customers and suppliers (it includes the financial response, which aims to compensate actors affected by the crisis situation and encourage the return of tourists, the crisis over); and.

b. the communication response, which aims to put in place measures to mitigate current and future losses.

Each of these responses requires actions. In operational terms alone, we can summarise the strategies available to tour operators to recover from a crisis as five:

a. shifting the offer from one receptive environment to another;

b. relaunching the product through a communication campaign ([31], p. 102);

c. market segmentation ([31], p. 102);

d. complete destination renewal; and

e. offer diversification.

For rapid and effective responses, these strategies need to be developed in advance of the disruption, bearing in mind that each crisis is unique and distinct from one another. Consequently, no matter how well thought-out, risk management responses will need to be revised during the crisis. In tourism, the aim is to shorten the lean period while working to eliminate the risk of destructuring. Based on the study of crises since 2000, the World Travel & Tourism Council and Global Rescue calculate an average recovery time of 11.5 months following a terrorist attack (2–42 months, depending on the scale of the attack); 16.2 months following natural disasters (1–93 months, depending on the scale) and 22.2 months following socio-political unrest (10–44.9 months, depending on the scale) ([32], pp. 5–9). For health crises, the average time is 19.4 months (ranging from 10 to 34.9 months, depending on severity), resulting in economic losses of 45–55 billion dollars ([32], pp. 5–9). In the case of the COVID-19 crisis, Gombault et al. ([15], 80) estimated in 2020 that it would take up to 3 years, or 2023, to return to "normality"—a prediction nearly right yet not completely fulfilled as of 2024. Hence, it is important for the tourism industry to be better prepared to withstand the blows, which would not be possible without the tourists themselves participating (financially) in this anticipation.

5. Responses from governments and the tourism industry

At the start of the COVID-19 pandemic, the sanitary measures adopted by many States (social distancing, disinfection of hands and contact areas, establishment of space partitions with Plexiglas, and most importantly, the wearing of face coverings) were primarily aimed at preventing the spread of the virus droplets during verbal communication. It was also a question of restoring the citizens' confidence to pursue their essential movements. Health authorities started by re-establishing a form of security in domestic life—grocery stores and other supply services, medical visits, schools, etc. Once tested, sanitary measures could then be extended to non-essential activities, such as tourism.

In order to resume their activities, the tourism industries needed to reconcile the conditions required for travel with the restrictions imposed (**Figure 1**). The relaxation of the sanitary protection measures recommended by the World Health Organisation from 2020 to 2022 was only possible if a transmission rate in the population of less than 1 was achieved, i.e. if one infected person contaminated less than one contact. And yet, the death toll inflicted by the virus—6 to 16% in G7 countries [33]—discriminated primarily against the elderly (aged 70 and over—a large segment of active tourists), but also younger retirees, those with a discretionary budget and time so essential to tourism. Hence, time became a significant part of the safety equation [34]. Herein laid the challenge of reactivating the economy, tourism and its industries, while maintaining social distancing measures. Should it fail, it would reignite the contagion, with a return to confinement to reduce the number of casualties.

Tourism recovery requires coordinated actions by several groups of players at all levels: political and administrative, local, national and international (opening of borders, creation and distribution of aid funds, financial support from the state and charities), insurance companies (coverage of pandemic-related risks, including care required in the event of contracting the disease while travelling, and, what is more, abroad), the medical sector (health care, capacity to respond to demand)

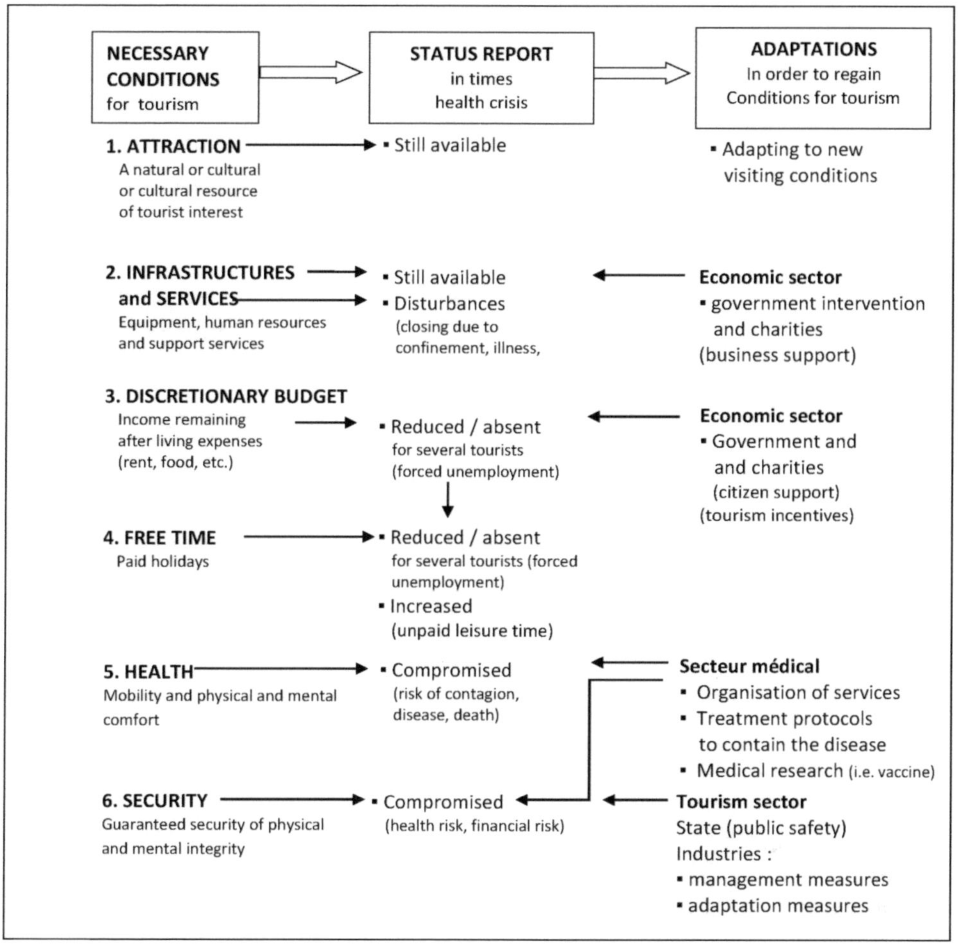

Figure 1.
Links to the impact of the COVID-19 pandemic on the six conditions necessary for tourism. Source: author's conception.

and, last but not least, the tourism industries themselves—transport, accommodation, catering and attractions, according to the contextual guidelines laid down by the public health authorities of both visitor-generating and host-receiving environments.

In the case of the 2020–2023 pandemic, the focus was lifting the lockdown, which was applied gradually according to the type of service and production method. In tourism, deconfinement was first allowed on a regional and national basis, according to the health status of host communities and tourist-generating regions, by type of activity (and age group). The same process was extended at the international level. The aim was to align deconfinement measures with the conditions set by the World Health Organisation (WHO) (**Figure 2**) [21]. Gradually, a number of tourism services and environments have been able to meet the World Health Organisation deconfinement requirements (taken over by national public safety authorities). But some key sectors faced significant obstacles due to:

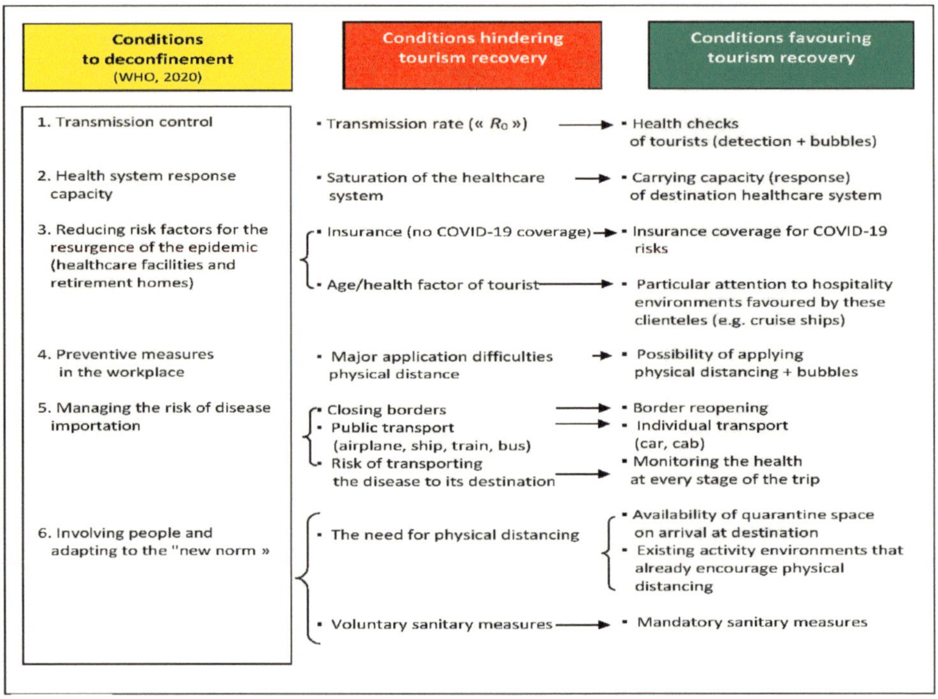

Figure 2.
Factors detrimental to tourism and conditions conducive to its recovery. Source: author's conception.

a. infrastructural considerations, notably with regard to space (cabin and passenger compartment) for collective mobility (aeroplane, train, coach and ship),

b. border closures or prohibitions on access to certain areas and territories, and.

c. the age and health of the clientele and host populations.

For tourists, safety is a question of perception—whether or not they are in danger (target) or exposed to danger (collateral victim). For example, territories afflicted by social or even armed conflict may continue to receive tourists if the crisis does not affect the area where holidaymakers are concentrated. In the case of the health crisis, no place could claim not to be at risk, since the tourist himself was the object of virus transmission.

The recovery scenarios considered by the various tourism industries during the pandemic were linked to the containment measures established by the governments. States' strategies differed according to whether they decided (a) to let things go without restrictive measures (as in the case of Russia, Brazil, the United Kingdom, Sweden, etc.) or (b) to partially lockdown the population and its economy (USA); or (c) to completely isolate and lockdown its borders while maintaining essential services (Canada, Québec, France, etc.). Two approaches emerged: mitigation (attenuation of the disease or its effects to make them tolerable) and eradication. The more aggressive approach to eradication has involved the rapid closure of borders, the rapid detection of the virus through mass screening and the use of technological applications (including sensitive data on the movement of citizens). These include Australia, China, Iceland, South Korea, Taiwan, Vietnam and New Zealand. What these countries have in common is that they are both geographically and politically insular. They have managed to control community transmission, enabling a return to normal life—at least, at certain times.

Three scenarios for the onset of a crisis then emerge:

1. no containment—business as usual, with no mandatory risk mitigation measures (and risks to the healthcare system);

2. complete containment—until a medical or pharmaceutical solution is discovered to protect the healthcare system, with known socio-economic consequences;

3. partial containment (managing the number of people in each other's presence)—with the introduction of infrastructural, temporal (curfew) and health risk management measures (implying that certain types of services and activities cannot be permitted).

It is possible to group together the measures deployed around the world as identified in the grey and scientific literature to mitigate the effects of the health crisis on tourism (**Table 1**). Hudson [18] provides an interesting overview with case studies from all around the globe, yet without synthesis. States and tourism industry players responded to the pandemic crisis first and foremost with information on the development of the crisis. Information and communication technologies (ICTs) then proved indispensable for managing socialisation in public places and enabling a maximum number of exchange networks to be maintained without physical proximity contact. At the same time, promotional campaigns on social media enabled the most imaginative destinations to remain "virtual" in people's minds.

In Iceland, for example, some tour operators have created wildlife excursions in order to broadcast live, via social media, commented experiences, in English, to an international audience (**Figure 3**). The interaction between customers and the tour operator, through exchanges of comments and the affixing of emotion icons (hearts, "likes"), enables the entrepreneur to hope to turn a few social media followers into prospective customers. Numerous destinations, particularly in film tourism and for museum exhibitions, used that approach.

Pandemic management required operational measures. In terms of health, these included screening for the virus and the disease, vaccination, application of sanitary measures to the general population, and certification of citizens' vaccination status as soon as the pharmaceutical response appeared. At the same time, governments deployed administrative measures to contain the virus, on the one hand, and to support households and businesses with financial aid, on the other. The tourism industry came to a virtual standstill, or had to rethink its functions—accommodation was sometimes used to relieve overburdened hospitals, or even as a depot for medical equipment. Some tourist accommodations offered their units for long-term healthcare, for employees who opted to telework from another location, provided inter-regional transport was available or permitted.

As soon as sanitary measures enabled economic activity to reopen, even partially, tourism was back in full swing. We then saw the deployment of new responses at both communication and operational levels (**Table 2**). In terms of communication, the

				Implementation authority		
				Governments	Tourism industry	Tourists
Communication responses	Information		• Need to keep the public informed	X		
	Use of ICTs		• Information and communication technologies (ICTs) to eliminate contacts – facial, voice, and fingerprint recognition (for the purposes of personal identification, contactless payment, registration, access to services via smartphones, tablets, room keys, etc.).		X	
			• Use of ICTs (paid videoconferencing) to present destinations or experiences.		X	
			• Installation of high-speed Internet services to promote working vacations at destination.	X	X	
	Promotion		• Promotional campaigns to keep the destination in the imagination of markets (notably via cinema-tourism)		X	
			• Use of virtual reality to promote destinations without travel		X	
			• Vouchers to postpone cancelled trips		X	
Operational responses	Health	Screening	• Health screening (temperature taking – by thermometer or infrared camera, nasal or buccal sampling)		X	
			• Medical monitoring by integrating medical teams into tourist accommodation companies.		X	
			• Telemedicine (telephone consultation)	X		
		Vaccination	• Mass vaccination of the population (from those most to least vulnerable to the disease)	X		
		Health measures	• Social distancing (widened with disease progression)		X	X
			• Wearing a face cover		X	X
			• Hand disinfection		X	
			• Surface disinfection		X	
			• Disinfection of shoe soles (some jurisdictions)		X	
			• Restricted socialisation (per bubble)	X		
			• Restriction on automobile traffic (time and number of passengers)	X		X
			• Curfew	X		

		Implementation authority		
		Governments	**Tourism industry**	**Tourists**
Certification	• Adoption of sanitary passes ("passport") with QR code (application by industries)	X	X	
Administrative measures	• Border closures	X		
	• VISA modification to encourage telecommuting tourists to stay (exempt from local taxes)	X		
	• Cooperation between neighbouring states to promote pandemic control	X		
Financial response (assistance) to companies	• State blitz survey to determine immediate needs	X		
	• Business aid/wage aid (in waves, for adjustments as needed)	X		
	• Loan interest rate relief/suspension	X		
	• Relief/suspension of loan payments	X		
	• Interest-free or low-interest loan/guarantee programme	X		
	• Business tax relief/suspension	X		
	• Tax relief/suspension to citizens on domestic travel purchases	X		
	• Salary reduction for employees		X	
Financial response (aid) for labour	• Employee/unemployed financial assistance (per cycle, for adjustments as needed)	X	X	
	• Economic support (for workforce training and retraining)	X		
	• Economic support (for training and requalification of workforce to other tourism jobs or employment sectors)	X		
	• Relief/suspension of contributions to social programmes (taxes)			
Service	• Change in the nature of service (takeaway food from restaurants rather than served in the dining room, end of self-service in catering, abandonment of tourist service in favour of other types of service, etc.).		X	
	• Change in clientele	X	X	
	• Policies to protect passengers in the event of transport cancellations		X	

Source: Compiled by the author.

Table 1.
Measures taken worldwide during the health crisis to mitigate its effects on tourism.

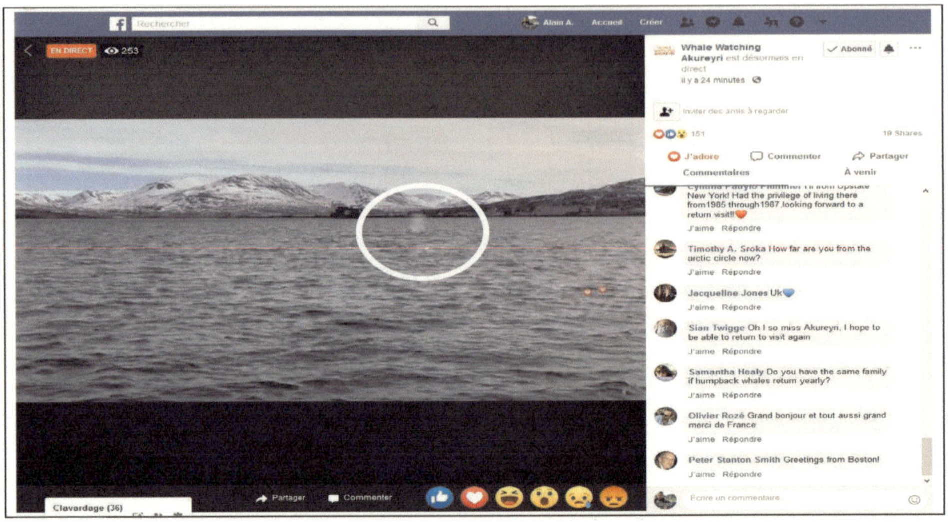

Figure 3.
Using social media to broadcast a live whale-watching tour (circle) to promote a tourism product to a prospective clientele around the world. Source: Screenshot taken by the author during a live session on Facebook.

strategies remained the same: keeping the public informed as the situation evolved and as to what was now permitted, or what restrictions were to be maintained. States had to play a balancing act in the face of grumbling, as certain freedoms were allowed while others remained prohibited, in terms of both socialisation and services.

Many destinations turned to the local market. This was the case with Norway [35] and Quebec [36], for example, two states which launched domestic marketing campaigns at the end of April and May 2020 to promote their offer and encourage local tourists to discover their own heritage (a solution that is both compensatory and adaptive). At the operational level, this implied new health safety measures. In the absence of medicine against the virus, the first summer, the Québec State encouraged people to rent cottages and bringing sufficient food supplies to prevent contamination of the local communities at their points of service. Many communities far from the main tourism centres would benefit from the influx of local tourists, often with unprecedented success, especially as vaccination allowed less restrictions. Day trips, otherwise known as "staycations", were also encouraged while waiting for the upturn.

With vaccines arriving in 2021, extending tourism to foreign markets required to obtain vaccination certification and flight or ship pre-departure temperature readings, adding to the administrative burden of the stay. International travel requires determination and patience. If tourism was to be revived smoothly, it was also necessary to restrict capacity (both in transport and at destination). Operators and hosts alike were required to review the layout of the space, keeping in mind the people's way of moving about inside buildings, disinfection of premises and even their bedroom quarantine for a few hours after customers have passed through. All these measures raised issues of manpower management, food supplies and profitability.

For destinations largely or exclusively dependent on international tourists (often sun destinations in emerging countries), the crisis continued. Based on data on the control of the spread of the virus, states established bubbles between privileged

			Implementation authority		
			Governments	Tourism industry	Tourists
Communication responses	Information	• Keeping the public informed	X		
	Use of ICTs	• Use of QR code for non-tactile services (via smartphone); (tickets, person ID, room keys, restaurant menus, etc.).		X	
		• Videoconferencing telephony		X	
	Promotion	• Change in appeal strategy (promotion)	X		
		• Use of different types of celebrities (artistic, sporting and influencers) to promote the destination.	X	X	
		• Special offers/discounts (free nights, single price for unlimited domestic transport for a fixed period)	X	X	
		• Focus on national/local clientele		X	
Operational responses	Security	• Adoption of regulations on visiting conditions (temperature screening of travellers).	X		
		• Adoption of "bubbles" and "corridors" between markets:	X		
		o depending on how they manage contagion			
		o by limited access: by air, land or sea			
		• Relaxation as quickly as possible of sanitary barriers to entry into a territory	X		
		• International recognition agreements on health passes (depending on vaccines)	X		
		• Imposition of travel insurance covering COVID-19 and its variants.	X		
		• Inclusion in travel packages of health insurance by destination.	X		
		• 24 h between the passage of guests in rooms in accommodation	X		
		• Imposition of quarantine periods on return from international travel (duration varies according to state and severity of epidemic)			
		• Cooperation between medial teams and hospitality services (accommodation)	X	X	
	Health measures	• Capacity restrictions		X	

		Implementation authority		
		Governments	**Tourism industry**	**Tourists**
	• Handling restrictions:		X	
	o withdrawal of cash payments		X	
	o disinfection by electrostatic gun and other ecological agents		X	
	• Vaccination and certifications (QR code passes)	X		
	• 72-hour screening (test before crossing an international border)	X		
	• Use of robots to provide certain services (site disinfection, meal delivery, etc.)		X	
	• Ban on mixing (foreign) tourists with the local population		X	
	• Maintenance of vacant rooms (in case of need to isolate tourists).		X	
	• 24 h between the passage of customers in rooms in accommodation		X	
	• insertion of Plexiglas dividers between user spaces		X	
Administrative	• Policies to protect passengers in the event of transport cancellations	X	X	
Financial response (assistance) to companies	• Fund to support creation and innovation (new tourism products/experiences)	X		
	• Workforce training fund	X		
	• Business recovery fund by sector	X		
	• Shifting weekend vacations to weekdays	X		
	• Preferential transport rates for local population (mobility incentive)	X		
Services and products	• Development of niche sectors favouring sanitary measures (physical distancing in particular) such as outdoor/nature tourism		X	

Source: Compiled by the author.

Table 2.
Measures taken around the world to boost tourism recovery.

markets (usually neighbours but also between states of different continents), allowing the most risk-tolerant tourists to reclaim air travel and cruises. The first cruises took place only along the coasts of one state, with customers from the same country and without stopovers. The upturn was marked here and there by cases of contagion, illustrating that taking the temperature of tourists before departure was no guarantee of a Covid-free trip [37]. These sporadic contagions required the tracing of travellers who may have been in contact, and their eventual isolation, applied with varying degrees of effectiveness depending on the state. Not all individuals—tourists and citizens alike—had the health of their fellow citizens at heart. At this stage, economic assistance programmes for businesses and citizens are generally extended. In some cases, rapid training programmes are added to help the workforce retrain for the job market once the crisis is over.

Not all attractions are equal in the face of a health crisis. In the case of the 2020–2023 pandemic, some tour operators showed originality by revising the formula of the experiences they offer (open-air cinema in nature parks, for example). Nature tourism gained in popularity as a self-sufficient activity. It often leads to overtourism, which has negative impacts on natural resource management. Where groups remained a necessity, and in emerging environments where medical or pharmaceutical treatment was inadequately supervised—the various vaccines on offer were not all recognised by the States, imposing other administrative barriers—some attractions and destinations suffered from the loss of most needed revenues.

6. Lessons learned

In the spring of 2023, the COVID-19 pandemic slowly faded away. As the World Health Organisation (WHO) lifted the health emergency on May 5, the world's economies began to breathe a sigh of relief [38]. The tourism industries are now drawing lessons from the reactions and adaptations developed—albeit often improvised—to counter the effects of a health crisis on a scale never before imagined.

These reactions and adaptations can be broken down into five general phases (normality, beginning of the crisis, during the crisis, end of the crisis) (**Figure 4**). Depending on whether the problem is non-recurring, recurring or permanent, various solutions emerge. In a state of "normality", operators of all kinds need to anticipate possible crises (those already experienced—recessions, socio-political crises, climatic and natural crises, cultural crises, criminal and terrorist crises—and now health crises). This means anticipating possible crisis scenarios and emergency plans, involving local and national authorities and the entire supply and support chain. The aim is to have a choice of possible reactions and actions in the event of the introduction of a disruptive agent into the socio-economic environment. The weakest link in any crisis management process is generally found upstream, insofar as the problem must be considered in advance by the stakeholders. However, small businesses often lack the resources, expertise, and time to do this, and they devote most of their efforts to the profitability of the moment.

The responses deployed by the tourism industries to the 2020–2023 health crisis can be broken down into three types of problems: non-recurring, recurring and permanent. A non-recurring problem generally occurs once or twice. It calls for short-term solutions. These are generally compensatory solutions in monetary terms (credit on the next visit, reimbursement or cancellation of activity). If they persist, the company must simultaneously suspend the product while promoting it, to keep it in

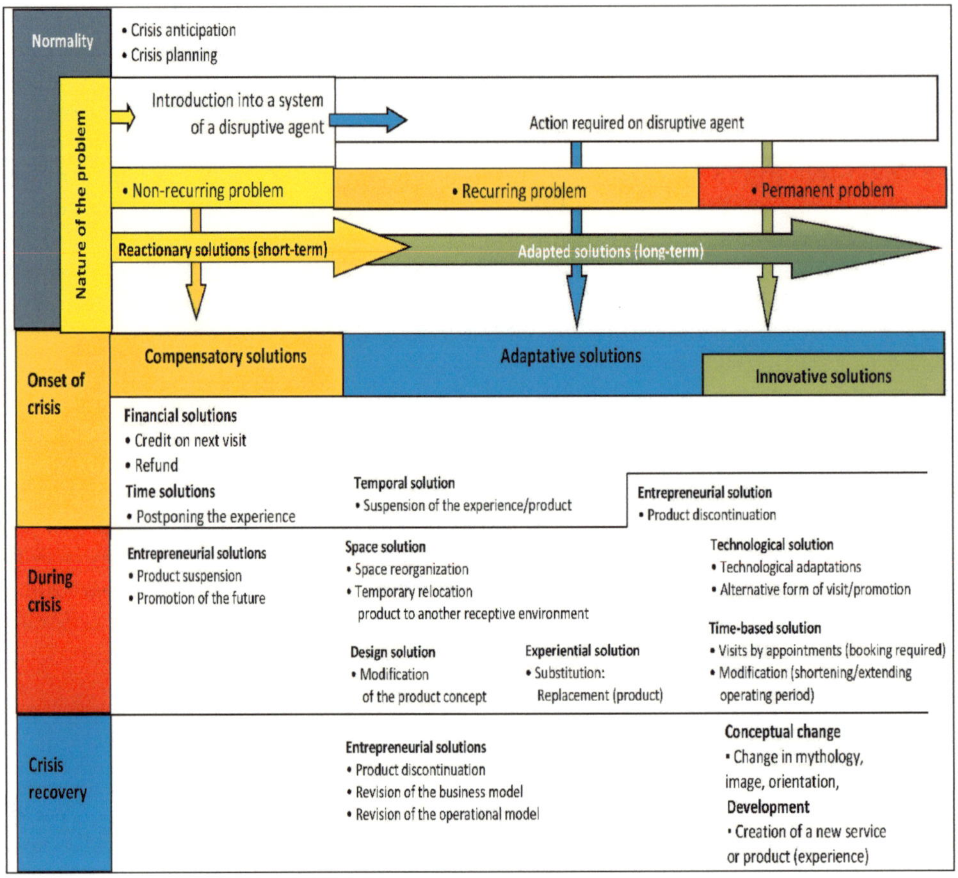

Figure 4.
The five possible reaction times and adaptations of tourism experience providers due to crises. Source: Author's conception.

the market's thoughts—a major lesson of the pandemic—and encourage its rapid relaunch when the crisis is over. Conversely, the product must be abandoned if the practice becomes permanently impossible in the long term (as is the case with climate change for many winter practices, for example).

When the problem becomes recurrent, or the crisis drags on too long (as in the case of pandemic 2020–2023), adaptive solutions (temporal, spatial, conceptual, experiential and entrepreneurial) are required. The temporal solution involves temporarily suspending the product or its experience. If the conditions necessary for its realisation are available by revising the organisation of the site, a physical reorganisation is required. The spatial solution may also involve moving the activity to another receptive environment to meet the criteria for safe execution. If necessary, the concept or form of the product or service can be revised. One experience can also be substituted for another.

In the event of health crises, pharmaceutical solutions may help restore normal functioning, enabling adaptive solutions to offer sufficient entrepreneurial and experiential resilience. In the case of disruptions that become permanent, the long-term effects of climate change are a case in point, and long-term solutions are required. These often involve innovation, i.e., improving a service or product by adding new elements. This can involve technological solutions (as during the COVID-19 pandemic, with visits to tourist sites via videoconferencing), temporal solutions (visits by appointment or by modifying service hours), conceptual solutions (by modifying the content of the experience, often in reference to its mythology, as in the case of climate change with snow) or, ultimately, by creating entirely new tourist experiences.

6.1 What did not work

Despite the many solutions put forward by governments and companies, there have also been failures—ideas that have not come to fruition. In general, governments intervened quickly to inject emergency funds to help companies survive. Certain sectors were initially excluded—air transport and cruising, only to receive aid in extremis (aviation only, as shipowners often register their vessels in tax havens). In the case of labour, the promised aid was often very uneven, even within the same jurisdiction, as in the case of smaller players—often individual or family businesses, whose cash flow is often extremely imitative. Since self-employed or seasonal workers are often less well-trained, they were the first to be laid off, finding themselves without recourse because they had not worked long enough. Without qualifications, it is difficult for them to work remotely ([18], p. 144).

In emerging countries, many companies and their employees live on daily earnings: without cash inflows, it is the end ([18], p. 146). Because of corruption and lack of trust in the central state, many citizens work outside the formal economy, escaping the tax system and, at the same time, state social programmes ([18], p. 74). In these situations, the pause in the economy and the social restrictions (mobility, proximity, etc.) introduced to halt the spread of the deadly virus could quickly lead to psychological distress, which in turn could lead to suicide. This shows how important it is for governments to combat corruption and abuse (unjustified expenditure, for example), which are partly responsible for the cynicism of citizens, leading to a shift away from the formal economy (and towards the informal economy).

Anti-vaccination campaigns, which were generally supported by people with little or no information, slowed down efforts to mitigate the pandemic. Sometimes, contradictory messages from the authorities contributed to misinformation. Pressure from the public to reopen the economy too quickly to foreign markets may have proved counter-productive when the relaxation of health safety rules made the destination a super-propagator.

The use of smartphones for a multitude of applications, thanks to QR codes, has encouraged “contactless” services. Already well established among young people under the age of 50, this technology has at the same time helped to isolate some of the elderly clientele, for whom this new way of interacting remains inaccessible due to sight problems, hand dexterity or other forms of a form of disability.

Another flaw was the neglect of environmental considerations in the aid programmes developed. Stopping tourism has had a positive impact on natural ecosystems in urban areas. However, in “wild” environments, the lack of income has affected the resources allocated to conservation, particularly in Africa ([18], p. 200).

7. Conclusion

From relatively localised ailments before the development of mass transportation, viruses are now increasingly called upon to disrupt economic and social activity on a globalised planet. There are thought to be at least 1.7 million unknown viruses in the world, of which 540,000 to 850,000 "have the capacity to infect humans." ([39], p. 5). Of these, 60 to 75% of new human infectious diseases (Ebola, Zika and influenza, AIDS, COVID-19) are zoonoses, i.e., derived from pathogens transmitted to humans by animals [40, 41]. Among the reservoirs most conducive to transfer to humans are mammals (bats, rodents, primates, birds)—especially seabirds—and farm animals (especially pigs, camels and poultry) ([39], p. 5), many of which come into contact with tourists as part of nature, agricultural or food tourism activities. From epidemic (local transmission) to pandemic (continental or global transmission), there is only one step: mobility.

The lessons learned from the 2020 to 2023 Covid pandemic can be divided into three main phases. Firstly, the world economy was brought to a temporary halt in order to prioritise the stabilisation of the health crisis. Even a partial resumption of tourism activity remained dependent on a variety of means that were not welcome in most cultures, such as physical distancing, behavioural (respiratory) hygiene, and physical and technological adaptations. The arrival of pharmaceutical solutions did not solve everything; instead, it brought much-needed relief to healthcare systems by mitigating the effects of the virus.

A second phase followed, known as partial cohabitation with the virus, in which strict rules were applied to mitigate or even reduce the potential for propagation in the populations the least at risk. When this second period was successful, the third phase could follow: a complete economic recovery, that is, a return to a new normality. This reality remains fragile insofar as globalisation continues to accelerate not only physical contact between humans but also original experiences with exotic fauna, which is also capable of transmitting viruses.

Despite the lessons learned from the pandemic, nothing is certain. The recent warning from the World Health Organisation [42] about the possibility of the H1N1 virus (avian flu) mutating into humans is a reminder of the vulnerability of human health to viruses and the strong possibility of another pandemic.

As each virus has its own specific characteristics, the adaptations developed during the COVID-19 pandemic are in no way a guarantee of the measures that will need to be put in place for the next health crisis. But the lessons of principle remain valid: the need for public health watch and health networks equipped in terms of manpower and operational capacities, the rapid reaction time of authorities at the first signs of crisis, and the quality of responses (support actions) from the population and businesses. This will require reserves to pay salaries and rents for several weeks or even months, which should directly appeal to tourists to pay more for their trip, something that conventional mass tourism, based on economies of scale, has yet to learn.

DOI: http://dx.doi.org/10.5772/intechopen.112775

Author details

Alain A. Grenier
Department of Urban Studies and Tourism, École des sciences de la gestion, Université du Québec À Montréal (ESG UQAM), Canada

*Address all correspondence to: grenier.alain-adrien@uqam.ca

References

[1] Statista. Number of International Tourist Arrivals Worldwide from 1950 to 2023. Statista. Available from: www.statista.com/statistics/209334/total-number-of-international-tourist-arrivals/; 2024 [Accessed: February 5, 2024]

[2] UNWTO – World Tourism Organization. 2023 Edition / International Tourism Highlights – The Impact of COVID-19 on Tourism (2020–2022). Madrid: World Tourism Organization; 2023. 32 p. Available from: https://static.poder360.com.br/2023/09/relatorio-Organizacao-Turismo-ONU-2022.pdf

[3] Stieb M. More Americans Have Now Died From the Coronavirus Than in the Vietnam War. Intenlligencer, 28 April. Available from: https://nymag.com/intelligencer/2020/04/coronavirus-has-killed-more-americans-than-the-vietnam-war.html; 2020 [Accessed: May 3, 2020]

[4] WTTC – World Travel & Tourism Council. Travel & Tourism Economic Impact 2022. United Kingdom: World Travel & Tourism Council; 2022. 36 p. Available from: https://wttc.org/Portals/0/Documents/Reports/2022/EIR2022-Global%20Trends.pdf

[5] WHO – World Health Organisation. WHO Characterizes COVID-19 as a Pandemic. World Health Organization, 11 March. Available from: https://www.who.int/emergencies/diseases/novel-coronavirus-2019/events-as-they-happen; 2020 [Accessed: May 17, 2020]

[6] Columbus C, Brust KB, Arroliga AC. 2019 novel coronavirus: An emerging global threat. Baylor University Medical Center Proceedings. 2020;**33**(2):209-212

[7] Lawrence SV. COVID-19 and China: A Chronology of Events (December 2019–January 2020). Washington: Congressional Research Services, US Congress; 2020. 55 p

[8] Siddiqui S, Alhamdi HWS, Alghamdi HA. Recent chronology of COVID-19 pandemic. Frontiers in Public Health. 2022;**10**. Available from: https://www.frontiersin.org/articles/10.3389/fpubh.2022.778037/full

[9] Crosby JC. Infectious diseases as ecological and historical phenomena, with special reference to the influenza pandemic of 1981-1919. In: Working Paper Presented at the Conference World System History and Global Environmental Change; 19-22 September; Lund, Sweden. 2003

[10] Gössling S. Tourism contribution to global environmental change: Space, energy, disease, water. In: Hall CM, Higham J, editors. Tourism, Recreation and Climate Change. London: Channel View Publications; 2005. pp. 286-300. Chapter 20

[11] Hall CM. Tourism, disease and global environmental change: The fourth transition? In: Gössling S, Hall CM, editors. Tourism and Global Environmental Change – Ecological, Social, Economic and Political Interrelationships. London: Routledge; 2006. pp. 159-179. 323 p. Chapter 9

[12] Kuo H-I, Chen C-C, Tseng W-C, Ju L-F, Huang B-W. Assessing impacts of SARS and Avian Flu on international tourism demand in Asia. Tourism Management. 2008;**29**(5):917-928

[13] Hall CM. Tourism and biological exchange and invasions: A missing

dimension in sustainable tourism? Tourism Recreation Research. 2014;**40** (1):81-94

[14] IFRC – International Federation of Red Cross and Red Crescent. The World is Not Ready for the Next Pandemic, Warns the IFRC, Press Release, 30 January. Geneva. Available from: https://www.ifrc.org/press-release/world-not-ready-next-pandemic-warns-ifrc: International Federation of Red Cross and Red Crescent; 2023 [Accessed: January 30, 2023]

[15] Gombault A, Lamarié J, Fouillet CG. La crise de la Covid-19, un changement stratgique pour le tourisme? L'analyse de la recherche internationale. Espaccs. 2020;**355**:79-85

[16] Walters G, McKercher B. Where to from here? COVID 19 and the future of tourism. Asia Pacific Journal of Tourism Research – Special Issue. 2021;**26**(11)

[17] Niewiadomski P. COVID-19: From temporary de-globalisation to a rediscovery of tourism? Tourism Geographies. 2020:651-656. DOI: 10.1080/14616688.2020.1757749

[18] Hudson S. International Case Studies on Tourism Destination Management and COVID-19: Impacts and Responses. London: Routledge; 2022. 240 p

[19] Blais M, Martineau S. L'analyse inductive générale : Description d'une démarche visant à donner un sens à des données brutes. Recherches Qualitatives. 2006;**26**:1-18

[20] Goeldner CR, Ritchie JRB. Tourism – Principles, Practices, Philosophies. Hoboken, New Jersey: John Wiley & Sons; 2009. 657 p

[21] WHO – World Health Organization. The 6 Steps – Six Criteria that each Country Should Meet before Lifting Restrictions. Thailand: World Health Organization; 2020. Available from: www.who.int/docs/default-source/searo/thailand/who-tha-six-steps.pdf

[22] Henley J. Just 7.3% of Stockholm had Covid-19 antibodies by end of April, study shows. The Guardian, 21 May. Available from: https://www.theguardian.com/world/2020/may/21/just-7-per-cent-of-stockholm-had-covid-19-antibodies-by-end-of-april-study-sweden-coronavirus?fbclid=IwAR2ac-i10GdZhzzx87xMd1lv_idKi1ekekI1FfKTMywOGsQZS6-WgNnNxK4. 2020 [Accessed: May 21, 2020]

[23] Rawaf S, Yamamoto HQ, Rawaf D. Unlocking town and cities: COVID-19 exit strategy, Eastern Mediterranean Health Journal. 2020. Available from: http://www.emro.who.int/in-press/commentaries/unlocking-towns-and-cities-covid-19-exit-strategy.html [Accessed: May 17, 2020]

[24] Grosjean, Arthur (2022) La 8e vague retombe – Les vagues Covid se succèdent toujours plus rapidement, Tribune de Genève, 2. Available from: www.Tdg.ch/les-vagues-covid-se-succedent-toujours-plus-rapidement-973545553291 [Accessed: February 5, 2024]

[25] Reuters. France's 8th Wave of COVID-19 is Gaining in Intensity, Health Official Says. Reuters, October 4. Available from: www.reuters.com/business/healthcare-pharmaceuticals/frances-8th-wave-covid-is-gaining-intensity-health-official-2022-10-04/; 2022 [Accessed: February 5, 2024]

[26] Küçükaltan D. Tourism and Terrorism – An Experience of Turkey and the World. Universe: Lincoln, Nebraska; 2006. 127 p

[27] Doane MA. Information, crisis, catastrophe. In: Mellencamp P, editor. Logics of Television: Essays in Cultural Criticism. London: Indiana University Press; 1990. pp. 222-239. 320 p

[28] Ritchie BW, Dorrell H, Miller D, Miller GA. Crisis communication and recovery for the tourism industry: Lessons from the 2001 foot and mouth diseases outbreak in the United Kingdom. Journal of Travel & Tourism Marketing. 2003;**15**(2/3):199-216

[29] Darling JR, Hannu O, Raimo N. Crisis management in international business: A case situation in decision making concerning trade with Russia. The Finnish Journal of Business Economics. 1996;**4**:12-15

[30] Grenier AA. Risque. In: Kadri B, Delaplace M, Grenier AA, Roche Y, editors. Vocabulaire du Discours Touristique. Québec: Presses de l'Université du Québec; 2022. pp. 399-403. 513 p

[31] Kanlayanasukho V, Veuthey P. Blood on the Beach – Managing the Impact of Political Crises in Tourism. Mereo Books: Gloucestershire, United Kingdom; 2016. 177 p

[32] WTTC and GR – World Travel & Tourism Council and Global Rescue. Crisis Readiness – Are you Prepared and Resilient to Safeguard your People & Destinations? UK: World Travel & Tourism Council and Global Rescue; 2019. 17 p

[33] Hopkins J. Mortality Analyses. John Hopkins University & Medicine – Coronavirus Resource Center. Available from: https://coronavirus.jhu.edu/data/mortality?fbclid=IwAR0fnp4dgsk7Y0 W32ISc4-M-alE4s OUnXZ8vS9V3 Z2PYRFZJpiPDZseJAg; 2020 [Accessed: May 22, 2020]

[34] Levenson E. Staying safe isn't just about hygiene and distance. It's about time, too. CNN, 18 May. Available from: https://www.cnn.com/2020/05/18/ us/coronavirus-time-risk/index.html. 2020 [Accessed: May 19, 2020]

[35] Local T. 'Norway is a Beautiful Country': PM Tells Norwegians to Stay Home this Summer. The Local, 15 May. Available from: https://www.thelocal.no/20200515/plan-for-a-norwegian-holiday-this-summer-pm-tells-people?fbclid=IwAR0nrorsLUI3BRk5jBbGv 8vzLujJ6h79LWO1VQ7HqJQk1muadj MoBlQxqKs; 2020 [Accessed: May 15, 2020]

[36] Rochefort A. Tourisme : une vidéo sur les plus beaux atouts du Québec destinée... aux Québécois. Société Radio-Canada, le 23 juin. Available from: https://ici.radio-canada.ca/nouvelle/1714445/campagne-bonjour-quebec-tourisme-video-david-goudreault; 2020 [Accessed: June 23, 2020]

[37] Street F. All People' Should Avoid Travel on Cruise Ships, CDC Warns. CN Travel, 23 November. Available from: https://www.cnn.com/travel/article/cruise-ship-cdc-warning/index.html; 2020 [Accessed: November 23, 2020]

[38] WHO – World Health Organization. Statement on the Fifteenth Meeting of the IHR (2005) Emergency Committee on the COVID-19 Pandemic. World Health Organization, May 5. Available from: www.who.int/news/item/05-05-2023-statement-on-the-fifteenth-meeting-of-the-international-health-regulatins-(2005)-emergency-committee-regarding-the-coronavirus-desease-(covid-19)-pandemic; 2023 [Accessed: May 5, 2023]

[39] IPBES – Intergovernmental Science-Policy Platform on Biodiversity and Ecosystem Services (2020) IPBES

Workshop on Biodiversity and Pandemics, Intergovernmental Science-Policy Platform on Biodiversity and Ecosystem Services, Germany. Available from: https://files.ipbes.net/ipbes-web-prod-public-files/2020-12/IPBES WorkshoponBiodiversityandPandemics Report_0.pdf [Accessed: March 6, 2020]

[40] Taylor LH, Latham SM, Woolhouse MEJ. Risk factors for human disease emergence. The Royal Society. 2001;**356**: 983-989

[41] Weiss RA, Sankara N. Emergence of epidemic diseases: Zoonoses and other origins. Faculty Reviews. 2022;**11**(2). DOI: 10.12703/r/11-2

[42] WHO – World Health Organization. Ongoing Avian Influenza Outbreaks in Animals Pose Risk to Humans, Statement, 12 July. Geneva, Paris, Rome. Available from:: World Health Organization; 2023 https://www.who.int/news/item/12-07-2023-ongoing-avian-influenza-outbreaks-in-animals-pose-risk-to-humans

Chapter 8

Strategic Management and Stakeholder Engagement: A Case for Environmental Sustainability

Ajay Khatter

Abstract

There has been an increase in industrial activity in the last few decades, but we must not lose sight of the importance of protecting the natural world. Industry stakeholders are concerned because we have yet to determine the long-term effects of this significant issue. Environmental sustainability is crucial given the world's increasing population, industrialization, and mounting evidence of climate change, environmental degradation, and global warming. Organizations have made significant efforts to reduce their environmental impact, and this study will explore the obstacles they must overcome to become environmentally sustainable. The study's findings may have important implications for businesses, clients, stakeholders, and the environment. The application of stakeholder theory has improved our understanding of company behavior, particularly in a dynamic and rapidly changing business environment, where environmental concerns are gaining traction. The purpose of expanding stakeholder theory could be to create fair value for all stakeholders while benefiting society and the environment.

Keywords: environment, sustainability, management, stakeholders, policy

1. Introduction

The modern industrial society has negatively impacted the global environment in numerous ways, most notably through air, soil, and water pollution, global warming, and climate change. This generation's defining issue may be climate change and its repercussions [1]. Consequently, it is acknowledged that the current rate of resource consumption cannot be maintained in the future. The greenhouse effect, which describes how carbon dioxide and other gases in the earth's atmosphere trap solar radiation in a greenhouse-like manner, is becoming more prevalent due to increased human activity, resulting in further global warming and climate change [2]. Governments are under increasing pressure from environmentalists, consumers, and communities to protect the environment [3]. This is essential for avoiding future harm, preventing irreversible damage, enhancing the ecosystem's capacity to support life, and allowing the regrowth of life forms whose survival has been threatened by excessive pollution and depletion of natural resources [4]. All parties must collaborate to manage and resolve a pressing issue like environmental sustainability. Stakeholder theory has been utilized frequently in sustainability management research to explain corporate environmental and social behavior and corporate social responsibility (CSR) motivations and drivers [5]. Understanding, developing an effective plan, and reaching a consensus are crucial components of stakeholder theory to achieve environmental sustainability. To increase the value of a concept—in this case, the idea of environmental sustainability in an organization—a variety of stakeholder perspectives and influences are required, as the best results and value cannot be achieved without their guidance [6]. Companies may miss valuable input, support, and advice if business partners are not involved as stakeholders [7]. This can be detrimental to the organization, as stakeholders' influence and power will impact the organization's initiatives. Engaging stakeholders in a company's decision-making process will help it achieve corporate responsibility and environmental sustainability over time, ensuring that they benefit from the results and contribute to their creation [7].

2. Literature review

In recent years, the global significance of environmental sustainability strategies as a core value has increased [8]. In recognizing and addressing their corporate social responsibilities and the environmental impact of their operations, organizational management needs to overcome significant obstacles. Brundtland Report defines sustainability as "development that meets present needs without compromising future generations' ability to meet their own needs" [9]. After quality management and information technology, environmental sustainability is anticipated to become a management priority of the utmost importance. The ability to implement sustainable practices will significantly impact the survival and success of an organization [3].

Sustainability encompasses the economic, social, and environmental aspects, although these have been treated separately to serve distinct interests. Social, economic, and environmental sustainability are interdependent, but in the developed world, social and economic sustainability has received more attention than environmental sustainability [1]. As a result of population growth driving increased economic activity to support social progress at the expense of the environment, it is still being determined whether environmental, social, and economic goals associated with sustainability can be harmonized [10]. There is a growing awareness that these three dimensions must work together because the earth's natural resources are limited, and the current generation must leave them to future generations.

Organizations have positively impacted the social dimension of sustainability by creating jobs, educating and training the workforce, enhancing working conditions, and supporting other social initiatives [11]. The increase in commercial activity has drawn attention to the economic aspect of sustainability. Consumption of goods and services has increased society's happiness, well-being, and social welfare, so this aspect has received the most attention. As a result, organizations have invested in infrastructure, manufacturing, plant, and machinery, achieving its primary objective of maximizing profits for owners and shareholders [12]. Economic activity has aided social sustainability by increasing employment and fostering a skilled labor force. Humanity has depleted ecological reserves and distributed waste that has contaminated the earth and its atmosphere to promote social and economic sustainability, impacting the delicate climate that sustains life. We are now experiencing the repercussions of failing to balance and coexist the three sustainability pillars [13, 14].

Failure to balance the three dimensions of sustainability leads to deforestation, biodiversity loss, resource depletion, the production of pollutants and long-term spoils, increased greenhouse gasses, world warming up, and harm to ecological surroundings [15]. We have long believed that environmental protection and economic growth are mutually exclusive. Environmental sustainability must precede economic and social sustainability if climate change and greenhouse gases have to be mitigated. These three components must collaborate and rely on one another to ensure sustainability. Moreover, achieving sustainability in these three areas is necessary for sustainability as a whole [10].

Internal barriers are one of the primary reasons organizations do not incorporate environmentally sustainable practices into their daily operations. Costs and a lack of legitimacy are internal barriers to environmentally sustainable practices, while regulation and industry-specific obstacles are examples of external barriers [16]. Small and medium-sized enterprises (SMEs) must be convinced about addressing environmental sustainability concerns. When attempting to address environmental management issues, SMEs face internal and external obstacles, but the latter is more crucial to their success [17]. Moreover, these companies are still determining the advantages of implementing environmentally friendly initiatives and practices. Due to a lack of client pressure, low client awareness and understanding of environmental sustainability, and other inadequate drivers, smaller organizations exert less effort to address environmental sustainability issues [17]. In addition, these organizations need help to locate and acquire high-quality environmental management guidance and data. Even when SMEs initiate and implement an Environmental Management System (EMS), the process is frequently halted, and resources are redirected to core business operations [18].

Freeman and Reed [19] developed the stakeholder, organizational management, and business ethics theory. This theory examines how morals and values should be applied to an organization's leadership for it to adhere to societal norms and standards. Freeman and Reed [19] developed the stakeholder theory, presenting ethical considerations based on the obligations and interests of stakeholders. According to stakeholder theory, an organization is a collection of stakeholders, and its objective should be to manage their interests, needs, and perspectives [20]. The stakeholder theory is predicated on the notion that businesses and organizations do not operate in a vacuum; instead, they rely on the assistance of various groups and individuals to generate value and facilitate trade. According to Freeman [21], stakeholders can be any group or individual who can influence or be influenced by achieving an organization's objectives. Included are parties with influence over environmental management policies and procedures. Employees, suppliers, future generations, customers, the government, shareholders, and communities are stakeholders in an organization [22].

To achieve sustainable development, resilience is a crucial factor that must be considered. To address sustainability, it is essential to be resilient and employ adaptive alternatives. Resilience may contribute to the reduction of uncertainty and unpredictability in sustainability management. For "resilience planning," stakeholders must work together because it provides a framework for developing new methods of collaboration and operation in response to the need for environmental sustainability [14]. Implementing environmental sustainability is difficult due to numerous stakeholders' diverse agendas and objectives. It may be difficult to strike a balance between the standards and operating policies of an organization and the financial goals of the owners and shareholders. All stakeholders must be informed of the difficulties posed by environmental degradation and why the organization should contribute to mitigating its impact [23].

The significance of the obstacles involved in achieving environmental sustainability indicates that achieving individual objectives is unlikely to resolve these obstacles. According to Hörisch, Freeman [22], stakeholder theory focuses primarily on individuals and groups who rely on an organization to help them achieve their objectives and upon whom the organization's survival depends. A prerequisite is the investigation and development of relationships between stakeholder interests that are mutually beneficial [24]. In contrast, stakeholder theory can support a broader objective of producing comparable value for all stakeholders to benefit society and the environment. To achieve this, the emphasis must shift from short-term financial gains or profit based on accounting to long-term societal and environmental benefits.

Significant and long-term change requires resources, the returns on which are frequently intangible. Businesses should recognize that investing in sustainability is a long-term strategy, even though certain practices and initiatives have short-term benefits. One of the most significant obstacles to environmental sustainability appears to be the expectation of quick or immediate returns on investment [3]. Using "strategy" in place of "initiative" and "investing" in the area of "spending" or "bearing costs" may result in a shift in environmental sustainability-related thinking and philosophies. Changing behavior may take decades, but the process must begin with a growing recognition that humanity faces a crisis of epic proportions [11]. Although the importance of environmental sustainability and the overexploitation of the planet's natural resources has been recognized for some time, business operations have yet to incorporate adequate initiatives to address these issues. Moreover, our environmental sustainability concerns are growing, and governments, world leaders, and organizations are attempting to address what was once a minor global environmental concern [1].

3. Discussion and recommendations

Educate stakeholders to encourage their participation in environmental sustainability collaboration. In a world preoccupied with the debate over climate change and the closely related environmental sustainability issue, stakeholders keep abreast of these as a focus of national governments and international organizations. To effectively inform and align all business stakeholders, organizational managers must establish a vision, values, and support practices in a formal written policy in consultation with all business stakeholders [12]. A sustainable culture can be created by integrating environmental sustainability throughout a company and appointing a leader for the initiative. Promoting a sustainable culture can also be aided by setting specific, measurable goals for sustainability initiatives. Working directly with owner stakeholders may make decision-makers, such as organizational management, uneasy, resulting in a radical transformation of operations and decision-making procedures [25]. This must be maintained, however, to obtain better inputs and support for addressing environmental sustainability issues. Developing and improving stakeholder relationships is one of the most crucial factors that can propel environmental initiatives and practices to succeed.

Despite the obstacles, balancing financial success and environmental sustainability is essential. To add value to the environment and society, stakeholders must

DOI: http://dx.doi.org/10.5772/intechopen.112065

supplement their short-term perspectives with a long-term view. Stakeholder theory remains open to the outcomes of stakeholder interactions, which necessitate a greater focus on environmental and societal contributions [26]. It should be noted that this shift in perspective and direction is likely to occur slowly in organizations, but as the importance and urgency of addressing environmental sustainability challenges increase, altering stakeholders' views may be one way to alleviate some of these pressures. To serve each other's interests and address environmental concerns, the emphasis must be on cooperation, finding areas of agreement, and gaining mutual benefits. If environmental sustainability is to have a more significant impact, stakeholder engagement must become a key component [27].

Having a formal policy on environmental sustainability ensures that employees have a direction for their daily responsibilities when senior management cannot be present. These guidelines improve an organization's internal effectiveness and goals, and objectives. Environmental sustainability policies demonstrate a company's initiative and foresight. The absence of a clear policy suggests ambiguous intent, while the presence of a policy suggests good intentions [28]. The mission statement, which describes the organization's objectives, should exemplify its environmental sustainability policy. An organization's policies establish a framework for delegating decision-making authority and defining its direction. Policies explain the consistent and transparent actions that organizations must take. The real value is only achieved when policies are implemented and closely monitored [28]. Identifying significant gaps between planned and actual implementation, as well as implementation enablers and impediments, is one method of policy monitoring. The recording and linking of policy changes facilitate the creation and implementation of future policies.

A formal policy on environmental sustainability suggests that this concept is separate from the organization's business philosophy. The most challenging aspect for these organizations may be approaching this crucial idea. Companies can initiate environmental sustainability by conducting research and developing a formal policy. The development of such a policy may be assisted by specialized environmental sustainability consultants retained by companies. This will help all involved parties formulate a clear strategy and comprehend what is required to contribute to environmental sustainability [12]. It will also demonstrate the management's aspiration to affect global change. Its written environmental sustainability policy and culture will reflect the organization's values. These will establish an ongoing commitment to environmental sustainability through formal policies and procedures that serve as the organization's guiding principles for education and guidance. If properly implemented and supported, these will strengthen the resolve of stakeholders to work together to promote and advance environmental sustainability. Implementing these policies will be facilitated by the guidelines' contribution to staff training and direction [29, 30]. Additionally, it can strengthen business ties between stakeholder groups.

Policies can be a potent communication tool for informing and bringing stakeholders together because all stakeholders can view them as a decision-making framework, concrete evidence of commitment and intended practices, and a component of the business strategy [12]. Although ad hoc sustainable practices must be implemented, formal policies can aid in defining the organization's objectives and ensuring that stakeholders cooperate and understand the significance of contributing to environmental sustainability. Environmental sustainability policies increase profits while simultaneously protecting the environment [31].

4. Conclusion

The environmental sustainability policy of a company demonstrates its dedication to environmental management. Without a policy, an organization lacks the necessary decision-making and implementation policies and guidance systems, making decision-making unpredictable [32]. The sustainability of the environment is one of the most pressing global issues. Recent years have witnessed a rise in concern for the health of our planet. Communities, nations, governments, and for-profit and not-for-profit organizations increasingly collaborate to address environmental issues. This has led to efforts to reduce our environmental impact and emphasize living harmoniously with nature. Scarcity of water, population growth, the need for more resources, climate change, and accumulating scientific evidence of environmental deterioration all contribute to the increasing significance of environmental sustainability [3]. To comprehend environmentally sustainable management's financial and brand-image organizational stakeholders must be aware of the benefits of going green. Stakeholders can encourage an organization's commitment to environmental sustainability by emphasizing its significance and acknowledging that they are already affected by environmental degradation. In the eyes of their stakeholders, organizations must bear a proportional share of the responsibility for ensuring environmental sustainability and improving the global standard of living, just as they do for their financial performance. Financially, organizations work to benefit their shareholders and investors, but they must now consider the environment an important stakeholder. Businesses that understand the numerous benefits of environmental sustainability, such as improved reputation and branding, competitive advantage, cost savings, increased productivity, reduced waste and resource consumption, branding as an employer of choice, and higher profits, will benefit significantly from going green. All parties involved—investors, shareholders, employees, customers, and the organization—win in this scenario.

Author details

Ajay Khatter
William Angliss Institute, Melbourne, Australia

*Address all correspondence to: ajay.khatter@angliss.edu.au

References

[1] Henson R. The Thinking Person's Guide to Climate Change. 2nd ed. Boston, MA: American Meteorological Society; 2019

[2] United Nations World Tourism Organization. Climate Change & Tourism. 2019. Available from: http://sdt.unwto.org/en/content/climate-change-tourism [Accessed: November 28, 2019]

[3] Rendtorff JD. Philosophy of Management and Sustainability: Rethinking Business Ethics and Social Responsibility in Sustainable Development. Bingley, England: Emerald Publishing Limited; 2019

[4] Lee SS, Kim Y, Roh T. Modified pyramid of CSR for corporate image and customer loyalty: Focusing on the moderating role of the CSR experience. Sustainability. 2019;**11**(17):4745-4767

[5] Serra-Cantallops A et al. Progress in research on CSR and the hotel industry (2006-2015). Cornell Hospitality Quarterly. 2018;**59**(1):15-38

[6] Schaltegger S, Burritt R, Petersen H. An Introduction to Corporate Environmental Management: Striving for Sustainability. New York, NY: Routledge; 2017

[7] Clarke A, MacDonald A. Outcomes to partners in multi-stakeholder cross-sector partnerships: A resource-based view. Business & Society. 2016;**58**(2):298-332

[8] Khatter A et al. Stakeholders' influence on environmental sustainability in the Australian hotel industry. Sustainability. 2021;**13**(3):1351-1366

[9] World Commission on Environment and Development. Our Common Future. Oxford, England: Oxford University Press; 1987

[10] Goudie AS. Human Impact on the Natural Environment. 8th ed. Hoboken, NJ: John Wiley & Sons; 2018

[11] Epstein MJ. Making Sustainability Work: Best Practices in Managing and Measuring Corporate Social, Environmental, and Economic Impacts. New York, NY: Routledge; 2018

[12] Kraft ME. Environmental Policy and Politics. New York, NY: Routledge; 2017

[13] Sarkis J, Zhu Q. Environmental sustainability and production: Taking the road less travelled. International Journal of Production Research. 2018;**56**(1-2):743-759

[14] Espiner S, Orchiston C, Higham J. Resilience and sustainability: A complementary relationship? Towards a practical conceptual model for the sustainability–resilience nexus in tourism. Journal of Sustainable Tourism. 2017;**25**(10):1385-1400

[15] Kihombo S et al. Is there a tradeoff between financial globalization, economic growth, and environmental sustainability? An advanced panel analysis. Environmental Science and Pollution Research. 2022;**29**:3983-3993

[16] Khatter A, White L, Pyke J, McGrath M. Barriers and drivers of environmental sustainability: Australian hotels. International Journal of Contemporary Hospitality Management. 2021;**33**(5):1830-1849

[17] Hoogendoorn B, Guerra D, van der Zwan P. What drives environmental

practices of SMEs? Small Business Economics. 2015;**44**(4):759-781

[18] Iturrioz C, Aragón C, Narvaiza L. How to foster shared innovation within SMEs' networks: Social capital and the role of intermediaries. European Management Journal. 2015;**33**(2):104-115

[19] Freeman RE, Reed D. Stockholders and stakeholders: A new perspective in corporate governance. California Management Review. 1983;**25**(3):88-106

[20] Mensah I. Stakeholder pressure and hotel environmental performance in Accra, Ghana. Management of Environmental Quality: An International Journal. 2014;**25**(2):227-243

[21] Freeman RE. Strategic Management: A Stakeholder Approach. Boston, MA: Pitman Publishing; 1984

[22] Hörisch J, Freeman RE, Schaltegger S. Applying stakeholder theory in sustainability management: Links, similarities, dissimilarities, and a conceptual framework. Organization & Environment. 2014;**27**(4):328-346

[23] Schaltegger S, Wagner M. Managing the Business Case for Sustainability: The Integration of Social, Environmental and Economic Performance. London, England: Routledge; 2017

[24] Freeman RE, Dmytriyev S. Corporate Social responsibility and stakeholder theory: Learning from each other. Symphonya. Emerging Issues in Management. 2017;**1**:7-15

[25] Davis K. Different stakeholder groups and their perceptions of project success. International Journal of Project Management. 2014;**32**(2):189-201

[26] Jang YJ, Zheng T, Bosselman R. Top managers' environmental values, leadership, and stakeholder engagement in promoting environmental sustainability in the restaurant industry. International Journal of Hospitality Management. 2017;**63**:101-111

[27] Crane A, Matten D. Business Ethics: A European Perspective: Managing Corporate Citizenship and Sustainability in the Age of Globalization. New York, NY: Oxford University Press; 2016

[28] Newig J, Koontz TM. Multi-level governance, policy implementation and participation: The EU's mandated participatory planning approach to implementing environmental policy. Journal of European Public Policy. 2014;**21**(2):248-267

[29] Yasin R, Huseynova A, Atif M. Green human resource management, a gateway to employer branding: Mediating role of corporate environmental sustainability and corporate social sustainability. Corporate Social Responsibility and Environmental Management. 2023;**30**(1):369-383

[30] Khatter A et al. Analysis of hotels' environmentally sustainable policies and practices. International Journal of Contemporary Hospitality Management. 2019;**31**(6):2394-2410

[31] González-Rodríguez MR et al. Hotels' corporate social responsibility practices, organizational culture, firm reputation, and performance. Journal of Sustainable Tourism. 2019;**27**(3):398-419

[32] Rodriguez I, Williams AM, Hall CM. Tourism innovation policy: Implementation and outcomes. Annals of Tourism Research. 2014;**49**:76-93

Chapter 9

Why Could the "Hatun Ñan Orqupi Llamcama" Ecotourism Microcircuit, Apurimac-Peru, be Sustainable?

Felipe Rafael Valle Díaz, Oscar Apaza Apaza, Rosmel Iván Rodriguez Peceros, Alfredo Huamán Cuya, Percy Lima Román and Artemio Montesinos Palomino

Abstract

The purpose was to describe and illustrate the characteristics that give sustainability to the use of the ecotourism microcircuit Hatun ñan orqupi Lamcama. The design is non-experimental, linked to the case study design. The sample was the closed circuit of ecotourism resources located in Llamcama's town. The method applied was on-site with the satellite capture support for orientation, location of resources, and mapping of trails, added to assess the vulnerability of ecotourism resources, use of the interview technique. Sustainability focuses on: (a) organizing local tour operators, who could serve a maximum of 80 visitors a day and 400 visitors a week, at a ticket cost of 5 PEN per visitor, total revenue per week and month, will be: 2000 PEN and 8000 PEN respectively; (b) accessibility is good to moderate in high season, changing to low in rainy season; (c) the security in the transfer and internal route is high, the signaling exists; d) the resources are characterized by being geometric shapes or anthropomorphic or zoomorphic approximation in the relief, caused by the forces of rain and wind; added to cave formations.

Keywords: accessibility, administration, inventory, tourist resource, tour operators

1. Introduction

We understand by ecotourism microcircuit, a small and controlled space with passageways or trails, made up of a series of natural structures, which were created by nature due to the action of rain, wind, and tectonic movements over time, and which in view of the human being, it is a great opportunity to take advantage of the landscape, reach mystical places to exchange energy, spending the night to admire the sky, and the noises of nocturnal species, among other attributes. Indicate that in the Apurímac region, there are comparative advantages, which the region in question has, supported by the ecological, geographical, water, cultural, and archeological aspects [1].

Ecotourism resources exist, they are categorized with level two or more, which are identified as: natural sites and cultural manifestations. However, there are limitations in the tourism sector, especially in the ecotourism type. The reason why tourist resources are categorized at level two, according to the guide for the inventory and update of tourist resources of [2] present little connectivity regarding technological opening of digital communication, moderate land access by trail or affirmed road that arrives even the tourist resource itself, local operators that are not very developed, dependent on the central support plant or main city, limited marketing development and positioning of the tourist resource.

The ecotourism resources are contained in the peasant communal territories, under the legal norm Official Gazette [3] Law 24,656 and [4] Law 24,657, the first known as: General Law of Peasant and native communities; the second Law of Demarcation and Titling of communal and native territories; what was verified and described by [5] indicates that the peasant communities, through the communal assembly, approve the organization of the protection and conservation work, based on the consensus of programming of annual activities, which are fulfilled and guarantee the conservation of the various tourist resources and natural areas or protected reserves.

The Pampachiri district, contained in the Apurímac region, has a vast ecological and cultural heritage, from the archeological, natural, and experiential, whose state of conservation is heterogeneous, for the reason that some resources are exposed and close to road access and others with road access limitations; Likewise, although there is progress in the way of exploiting the ecotourism microcircuit through the formation of tourist committees (association), these present a series of difficulties, which are focused on customer service, the conformity of the service provided, the capacity to respond to unforeseen events, security during the internal journey, among others.

The microcircuit that exists, has the minimum services necessary to continue sustaining itself in a precarious way when providing the service. The peasant community has a certain level of interference regarding the control of these tourist resources, so vulnerability is low, but progressing, because it is subject to the way in which ecotourism resources are used. The problem is centered, on the existence of deficiencies in updating the inventory and in the way in which tourist resources are used, it is a space or area that has roads and trails that allow limited transit of vehicles, limitations in spaces or collection points, and waste evacuation, some deficiencies in control and security during the internal route, limitations in the existence of signage.

The general question of the investigation was: Will it be possible to take advantage of the ecotourism microcircuit "Hatun ñan orqupi Llamcama," district of Pampachiri? This resulted in formulating the following general objective: Describe and illustrate the characteristics that give sustainability to the use of the eco-tourism microcircuit Hatun ñan orqupi Pampachiri. The dimensions that built the research were: (a) the organization of local tour operators; (b) accessibility to and within the microcircuit; (c) security in the transfer, internal route, and signaling; and (d) the attributes of ecotourism resources.

The justification of the research project focused on two aspects: the legal-academic relevance, and the impact of the proposal for sustainability:

a. Regarding the first, it is located within the UNESCO research area, area number 53, sub-area 53.11, research line 53.11.06 market study; within the Regional Research Agenda approved with Resolution No. 013-2015-GR-APURÍMAC-CR, economic agenda, tourism component, rural-community and archeological tourism thematic area, and line of research on the subject: Design of insertion processes and articulation of corridors and tourism products in new areas with tourism potential.

b. The second, from the socio-economic point of view, in the medium term the ecotourism microcircuit will help to improve income and position the cities of Pampachiri and Andahuaylas, as cities with ecological, historical, and touristic potential of cultural. In addition, it will materialize with the increase in demand by foreign and national tourists, between the short and medium term.

The scope of the investigation is an extension of the first stage that focused on the San José de Chontaya annex of the same district, and a specification of the second stage, aimed at describing and strengthening the use with the difference of proposing the conditioning of tour operators, both located in the Llamcama Town Center.

The concepts of accessible tourism are taken, linked to the theory of alternative tourism, which contains ecotourism, one of them being what was stated by [6] who indicates that "accessible tourism allows people with access requirements, in mobility dimensions, vision, hearing and cognition, function independently with equity, dignity, through the delivery of universally designed tourism products, services and services," (p. 519) in the concept of [7] who express that: "accessible tourism it is an activity that all people can access regardless of their abilities, thus enabling full participation and integration in society as a whole." (p. 26) We close the concept stage, We close the concepts stage with [8], who specifies that: "accessible tourism must be understood from a global point of view, covering the set of facilities and operators that make up the tourist offer, delving into entertainment and leisure services, among others" (p. 59).

Based on what was established in the concepts of Accessible Tourism, the same background of similar investigations was continued, to strengthen the methodology, design, and comparison of results, these being:

Being the contribution of [9] they mentioned that tourism is an alternative to improve the quality of life of households in communities. Creating tourist routes is essential because they allow for expanding tourist activity, however; the final results depend on various actors, involvement, and coordination are essential. The investigation concludes that the tourist routes located in the province called "El Oro," are not properly exploited, problems in tour operators, and the conditioning of tourist infrastructure by the local municipality. The methodology used was: documentary analysis, a historical-logical method, complemented with observation and interview.

Another perspective, by [10] states that protected natural areas and parks present benefits for environmental conservation; However, it offers opportunities for outdoor recreation and for the enjoyment-awareness of preserving the landscape, both associated, because they are linked by a variety of scales of improvement and economic use, such as the case of the socio-ecological system in which all resources interact, facilitating an ecotourism articulated with the support plant.

Public parks fulfill the function of strengthening ecological tourism, allowing biodiversity to be displayed and preserved, preserving small established ecosystems, allowing sustainability; However, for the case study, the results show little sustainability for the settlements, there is less development of marketing and advertising tactics; The final role of the state should be the design of a project to consolidate and recover nearby tourist resources, through the promotion of investment packages [11].

We have [12] from the investigation of the design of the ecotourism route in Coloso-Sucre; It focused first on identifying the obstacles related to the infrastructure of the tourism plant, the weak tourism promotion by the local municipality. He described the natural, historical, and cultural potential that Coloso-Sucre possesses; and concludes by proposing a descriptive design of the tour based on natural attractions, history, customs, architecture, culture, and involving tourism service providers, with the promotion of the local municipality.

From [13] they indicated that the local municipality from 2006 henceforth, strengthened the creation of an ecotourism center as part of a development policy, based on the commercialization and promotion of rural tourism, influencing investment in infrastructure improvement and training to local residents, incorporating them into the economic model, as a lifestyle to increase their income, through the diversification of some productive activities, sustained businesses in kinship ties and geographical space.

Studies [4, 5] show us that peasant organizations are a semi-commercial economy, with decapitalized resources, frequent use of available eco-tourism resources, with limited projection regarding improvements or renewal, which do not allow the leap of growth, horizontal, as well as limitations in peasant education and organization. According to [6] the assets owned, are usufruct based on some characteristics made by [14] the perception of local communities about the development of ecotourism, it focuses on: (a) preserving cultural traditions, (b) sustainable community, and (c) planning and ecotourism management with state intervention. The results show a significant association between the attitude and the intention to preserve the cultural tradition with the planning and management of ecotourism.

According to [15], they influence that to expand and strengthen the tourist practice, first promote it, based on the implementation of a didactic medium as a tool, in such a way that good practices with the environment and sustainable use are disseminated; The results showed that over 50% are aware of what ecotourism is and 37% have difficulty understanding ecotourism.

The contributions of the authors, on the ecotourism route of the village of Shitabamba in the national park of Cutervo, Cajamarca [16]. The design focused on the active participation of the community, who formulated, based on their knowledge of the territory, the potential in tourism resources they possess. The methodology was based on unstructured interviews, complemented by the inventory methodology of the Ministry of Foreign Trade and Tourism. The research project provided the justification for the formulation of the Cutervo National Park.

Ref. [17] stated that the Canton of Tena has the potential for tourist resources, but that they do not have the conditions for development toward a tourist attraction, due to the absence of adequate infrastructure, which affects the well-being of the population, impediments to recover and strengthen culture, added to deficiencies in the population; the study proposes the design of a minor circuit, based on local businesses, with the focus of preserving the natural environment [17].

2. Methodology

This research has a qualitative approach, due to the characteristics of the study variable according to [18]; of the basic type, which allows us to provide current knowledge of the study variable from [19]. The design is non-experimental, linked to the case study design, because it analyzes the diagnoses, the characteristics, the accessibility, and the state of conservation of the ecotourism microcircuit. The description allowed us to propose modifications, and adjustments both in the microcircuit and in the planning of tour operators, this is supported by [18].

The population was the geomorphological and cultural system that contains the territory of the Populated Center of Llamcama, and the ecotourism resources sample, as exposed in [18]. The support to collect information was the inventory guide and update of tourism resources of the Ministry of Foreign Trade and Tourism of Peru, approved in December 2018.

The data collection procedure was as follows: first, field trip for coordination with local authorities and signing of commitment documents setting a start date; second, induction talk with participants about the work that has been programmed to be carried out under the policy and guidelines on research of the José María Arguedas National University; third, training talk and review of the process in the field, carried out with the local authorities and the community organization; fourth, information gathering with the support of local authorities, for each tourist resource, the impression of satellite captures was used to identify the tourist resources, then on the Google earth platform measurements were made of the surface it occupied, likewise, we measure the distances with this tool and the support of GPS, we continue with the capture of photographs of flora and fauna species, we close with the data collection with a check sheet regarding the services provided and their level of quality; fifth, collection and summary of secondary sources to strengthen the interpretation and description, and sixth, systematization in the office.

3. Results

3.1 The organization of local tour operators

There are no hostels, lodgings, or restaurants of level or category from 2 to 3, in the indicated neighborhoods of the populated center. For this reason, tourists or visitors who arrive at Llamcama must choose three alternatives:

a. The first alternative, is to spend the night in the district capital of Pampachiri, where there is a lodging infrastructure, and hostels, in regular to good conditions and even with complementary services, this will allow the tourist to rest easy, and can leave the next day, without the long journey from the city of Andahuaylas. There are restaurants, whose service is minimal because there is no variety in the gastronomic offer and restricted access to toilet service. There are no night games, or shows in the district capital Pampachiri; however, the interview with owners indicates that it can be implemented in lodgings and hostels.

b. The second alternative is to stay overnight in Llamcama, this is possible if they agree with the authority to provide a local community or environment of the local municipality (saves lodging costs) followed by coordinating and agreeing on a food business with families known that provides the food service; It has minimum basic conditions for national or international visitors of the backpacker type, the need for overnight stays, is based on providing communal premises, which are conditioned with mattresses and blankets. Communal premises, in general, have several separate environments, and hygienic services in a regular state.

c. The third alternative is to return to the city of Andahuaylas the first day, spend the night, and leave the next day at dawn to arrive at 9:30 am and continue with visits to tourist resources; this is tiring, but it can be tolerated by visitors who are not located in the third age or older adults; This will have to be done at least twice, at most three times. The city of Andahuaylas, has sufficient tourist infrastructure, varied, as well as places for recreation. Added to the fact that the land terminal and transport companies are located.

On the use of tourist resources by tour operators, they have implemented and conditioned the following: In agreement with the Directorate of Foreign Trade and Tourism, added to training, they installed the tourist ticket, which is executed as a statistical registration procedure and attention in two entrances: Huaccoto and Llamcama. At the moment the administration of the ecotourism microcircuit has two types of administration, to reduce the degree of conflict; the entrance of Huaccoto, exercises a first control and registration; and the Municipality of Llamcama the second control and registration.

The first entrance has a higher frequency of visitors compared to the second. The management or attention regarding the tourist orientation that is provided to know the attributes of the tourist microcircuit, focuses on the number of members per group per orientation service, which is a maximum of 7 to 8 members, who can be easily guided. This allows for a maximum average of 80 visitors per day and 400 visitors per week. If the cost of admission is 5.00 soles, the total income per week reaches 2000.00 soles (two thousand PENs) and per month a total income of 8000.00 (eight thousand PENs) will be obtained, this is achieved as a goal in high season between the months of May to October, because the low or rainy season is from December to April, and visits are reduced due to the inaccessibility of the carriage trails.

A social problem that is addressed by the Municipality of Llamcama, is to give guarantees and formalize giving job opportunities to mothers with dysfunctional families; abandoned, which is justified because of every 5 women between 1 and 2 present this difficulty in generating income for the family, this is done through a resolution act and delivery of identification clothing (vests) they are granted authorization to develop tourist orientation, previous training with tutelary institution; In addition, this singular group is immersed in earning a living in various activities. They are the ones who presented the greatest predisposition to start in the tourist activity; Likewise, these operators in tourist orientation, are the ones that present communication between the Municipality of Llamcama, and the Primary Care Health Center.

3.2 Accessibility to and within the microcircuit

To get to Llamcama, you start from the city of Andahuaylas, the section covers 114 km and a travel time of 3.20 hours, 85% of the section is paved national highway, then you enter by variant, affirmed road that includes the 15% of stretch. Llamcama is located between 25 to 30 minutes from Pampachiri. Access to ecotourism sites is regular, through defined paths, for people with full motor skills, and for those who have them partially, horseback riding can be provided. People with complete disabilities or invalids cannot take the services. To reach the archeological resources, they do not have adequate accessibility, and the paths in a regular state on a steep slope, with a long extension and pronounced fatigue, they allow you to reach the inaccessible peaks where the archeological resources are located. This tour can only be done by visitors with full motor skills and a state of health without respiratory or cardiac problems. The visit to the ecotourism circuit can be done as follows:

The first day, the closed circuit that starts from Llamcama, the ascent section is not very rough, we start in a southerly direction, we arrive at Wasa millpu, a cave that at the entrance is surrounded by a forest of cceñuas and outside the cave, the forest indicated the it surrounds, in addition, there are several entrances to enter the interior and start an internal walk. Outside the cavern there is a location signage in a poor state; Inside, it has conditioning works for limited access due to the conditions of the cave itself, there is no lighting to be able to continue beyond 150 linear meters (**Figures 1** and **2**).

Figurc 1.
Forest of cceñuas. Forest of cceñua species, covers about four hectares, is bordered and inside there are spaces for camping, located in the Millpu sector, 30 minutes by car from the town of Llamcama.

Figure 2.
Chinkana Wasa Millpu. Nota. Entrance to the Wasa Millpu cavern, it has a length of more tan 160 meters, with steep slopes inside, it forks into two entrances, both explored and in the process of improvement, located 30 minutes by car from the town of Llamcama.

Then we continue north, leaving for Huanca Raza or the place where the condors perch, this ecotourism resource can be seen from the base or can be climbed, allowing both positions to take pictures; Climbing to the summit is not very accessible, only people with complete motor skills can do it. When we reach the summit, we are 4300

Figure 3.
Huanca Raza. Nota. Close view of Hatun Huanca Raza, or known as the place where the condors rest and eat, has a single rugged access to the summit in a northeasterly direction, with a steep slope between 25° to 35°, located 20 minutes by car from the town of Llamcama.

meters above sea level, it serves as a natural viewpoint; up to here, we have 3 hours of travel (**Figure 3**).

Then we continue in a northerly direction until we reach the signposted entrance to Chinkana Wisccacha, we walk up the marked path along the edges, the slope is on average 15° to 25° degrees, and the travel time to Chinkana Wisccacha is more than 20 minutes the almost 70 linear meters; When entering the chinkana, (cave) the path is narrow in sections, the rock is sculpted by the force of water and wind, it presents different shapes, the route is gently ascending, there is a presence of strong humidity in the environment, the time does not exceed 25 minutes, disabled people of any lower or upper limb cannot enter the cave, added that the elderly group over 60 years, cannot enter due to the degree of average humidity in the caves, which would bring respiratory and even pulmonary consequences, complements that in the internal route, there is always a watercourse (**Figure 4**).

Then we continue north toward the signposted resource of Awekillaqta, we travel about 25 minutes, to appreciate the terraces, protection, and shelter houses carved in the stone, with the presence of terraces. So far, we have 2 hours of travel and enjoy the landscape. Continuing, at the exit of the circuit we reach the first stone forest which can be reached in 10 minutes.

On the second day, we started from the town center of Llamcama, taking a shortcut to the east until we reached the Awekillaqta tourist resort, we continued on our way behind this tourist resort to move toward the Apu Ataccara and Cemetery, in a northeasterly direction, with respect to the Cemetery this Located at the foot of the indicated hill, which takes 10 minutes to arrive from Awekillaqta, among the attributes we find funerary constructions from different times, decoration styles based on culture. We continue, on foot, climbing the Apu Ataccara, bordering it, halfway up the slope and reaching archeological resources located at the summit, which takes an hour and a half, in addition to see the landscape, at the summit we stop for half an hour, up to here we have 2 hours and a half of travel. We continue southwest, first, we descend and then we ascend on foot toward

Figure 4.
Chinkana Wisccacha. Nota. Close view of the entrance below to the Wisccacha cavern, it has two accesses which are used for the internal walk, from top to bottom and viceversa, with a steep slope between 15° to 25°, located 5 minutes by car from the town of Llamcama.

the Piedra Corazón tourist resource, reached in 45 minutes, which is a lithic form, carved by the action of nature; wind and water. We return following the bed of the Llamcama stream, in a northeasterly direction, to reach the outcrop of Ccoñecc hot springs, near the town of Llamcama. The entire tour is 4 hours on the second day (**Figure 5**).

Figure 5.
Site Awekillaqta nota. Close view of the Awekillaqta sire, it has only one access, the road allows máximum access, with a steep slope between 15° to 25°, it contains buildings that were supported on rock formations, whose spaces inside were inhabited with limestone, outside the terraces on three levels, it is a site of several constructions, located 10 minutes by car from the town of Llamcama.

3.3 Security in the transfer, internal route, and signaling

The Llamcama ecotourism microcircuit offers a uniform transfer by the main national road, marked in several sections, with support points. Inside the resource, the trails and internal trails are conditioned with material from the area, it also has the installation of rest or rest points for visitors and tourists, it presents limitations in the conditioning of a waste classification collection or evacuation system Solid, 50% of the edges of the trails are signposted, with white stones embedded in the soil. Regarding the care and protection of the various ecotourism resources, there is progress as proof, it is the Awekillaqta resource, with a perimeter protection fence. The entrances have a control booth, which fulfills the function of a ticket office and the municipality of Llamcama, has a record book. The groups of visitors will be able to choose the package that is offered, these being:

a. Tourist package "A": Wasa Millpu Cave (Chinkana), Apu Huanca Raza, Wisccacha Cave (Chinkana), Awekillaqta, Stone Forest; from 9:00 a.m. to 2:00 p.m.

b. Tourist package "B: Cemetery, Apu Ataccara, Heart Stone, Ccoñecc Hot Springs, from 9:00 am to 1:00 pm.

Within the security protocols during the trails, they present the following service and prevention measures: all community tour guides or tour operators must follow the established and marked path or path, accompanying the group, the community tour guide, that he has trained in guiding people. The community tourist guide knows about history, and geography for a better interpretation and also about first aid, for unforeseen

Figure 6.
Counselor and citizen security in microcircuit. Nota. Counselor and citizen security – Irma – at the Awkillaqta site by the municipality of Llamcama.

situations that may occur regarding the type of ecotourism route where the service is offered. The association of community tour operators was established based on the support of the Pampachiri District Municipality and the Llamcama Town Center, the Association of Llamcama Tour Operators (**Figure 6**).

3.4 The attributes of ecotourism resources

We find the natural sites called:

a. Ccoñecc hot springs, which is an outbreak of thermal water south of the town of Llamcama, the flow level is permanent, the temperature is maintained in parameters of 20° to 30° degrees Celsius;

b. Chinkana Wisccacha, a natural gorge located in the southwest direction, with a length between 60 to 70 linear meters whose route is zigzagging, ascends, and descends, the Stone Forest.

c. In addition to cultural manifestations such as: first the Apu Ataccara, with archeological constructions, second the Awekillaqta archeological remains, Inca roads or carriage, intact with fences, and speed bumps, added to the native culture that is located in the Llamcama annex.

All the ecotourism attractions were formed by actions of nature, with the exception of the archeological zones, which are in the process of deterioration due to the interference of man and nature factors. In the archeological zone, the constructions have a circular shape with an inclination toward the interior, they settle stones on stones in some cases with clay or mortar with inputs from the place, and at the base large stones; each house is contiguous to a corral or space destined for another function that forms small groups of human settlements, there are also internal paths, which demonstrate the order of human settlement, added to the anthropomorphic evidence that exists in the Apu Ataccara.

4. Discussion

The recorded background, as a reference basis for the process and similar methodology, allows us to compare, and approximate results, in addition to taking considerations for the improvement of tourist resources, subject to current research, Llamcama Population Center is considered in this discussion, as it contains the units of analysis:

For the case study, we have continued to describe the potential, making an initial inventory, followed by marking the route or direction of the trail, travel time, and accessibility conditions of the ecotourism microcircuit. To complete this first stage of work, we have carried out a study of the relations of the population with the tourist resources, which are necessary because they will provide support and support to any project initiative. The state of the nearby support plant and the main city of the tourist resource have minimal services, which gradually improve in response to the progressive demand of visitors and tourists.

In addition, in consideration for the investigation, we have continued to describe the current way in which they make use of the ecotourism microcircuit according to the rainy season, which facilitates access. The gradual use, under minimum standards of exploitation and adequacy of the ecotourism microcircuit, generates sustainability, from the implementation of a strategy of administrative control, collection, and recycling of solid waste with an incipient system implemented by the organization of tour operators.

The visitor carrying capacity to minimize: noise pollution, species stress, and landscape deterioration is a maximum of 40 visitors per day. The relationships and interests of residents and tour operators with the tourist resource, add the need to maintain the use based on the exploitation of two internal routes, which are marked with precision to 40%. It should be noted that there is an organization in the neighborhoods that allows programming the use of the ecotourism microcircuit; the problems of territorial boundaries between the neighborhoods contained within the communal territory, were overcome.

The municipality has started a capacity building program, and is even in the process of designing investment projects to start physical-legal sanitation and protection facilities for ecotourism resources.

5. Conclusions

The potential of tourist resources contained in the Llamcama ecotourism microcircuit is diverse, only two internal routes are being used; Sustainability would be based on the administration of the complex by the association of community tour operators, which will allow establishing controls, and monitoring so that the resources contained in the ecotourism microcircuit do not deteriorate. The peasant communal organization is where the Association of Community Tour Operators (AOTC) was born. The AOTC will be in charge of managing, protecting the rational exploitation, and maintaining the trails. From the environmental and cultural point of view, it will contribute to consolidating an area of protection and use of ecotourism and archeological resources.

The local Municipality will tender the closed ecotourism microcircuit to the association of tour operators prior to the bankruptcy process, who must develop the registration and control at the entrance and inside. One of the economic resources that will finance maintenance will be collecting a percentage of the total cost for each ticket sold. The distribution of the entry ticket, equitable, will be under the proposal: 1.00 soles (one PEN) for the security guard or casetero, 1.50 soles (one PEN and a half) for the association, and 2.50 soles (two PEN and a half) for the Municipality of the Center Town, which assumes the cost of issuing the ticket.

From the environmental point of view, the way of managing the visits allows for a low impact due to human intervention, in the conditions that exist such as: landscape, acoustics, solid and liquid waste, and stress of animals and plants. The closed ecotourism microcircuit of the Poblado Menor de Llamcama center has the following attributes: Chingana, Awekillaqta, Apu pavilions, Stone Forest, Red Apu, Pankula, Heart Stone, Ccoñecc thermal water, Suytu Machay caves and the Llamcama canyon. There is a moderate to low degree of vulnerability of various tourist resources, due to the disorganization in the administration, maintenance, and conservation of ecotourism resources.

The accessibility is good to moderate in the high season (there is no rain), changing to low accessibility in the rainy season. The support plant close to the tourist resources presents limitations in various services, which are complemented by a support plant in the district capital and the city of Andahuaylas. The carrying capacity is an average of 80 visitors a day and 400 visitors a week, at a cost of 5 PEN per visitor, the total income per week and month will be: 2000 PEN and 8000 PEN respectively.

The District Municipality of Llamcama will have to lead the formalization of the association of community tour operators where priority must be given. Likewise, it will have the duty to implement the protection of tourist resources up to 75% to stop the violation and deterioration of ecotourism resources, likewise, it will implement spaces with solid waste sorting bins and hygienic services in strategic places.

Acknowledgements

Felipe Rafael Valle Díaz, contribution to the recognition in the publication: conceptualization, data curation, writing-review and editing of the published work.

Oscar Apaza Apaza, contribution to recognition in publication: methodology, verification of data systematization.

Rosmel Iván Rodríguez Peceros, Percy Lima Román, contribution for recognition in the publication: conducting research-data collection, supervision and leadership to carry out activities, initial draft writing.

Alfredo Huamán Cuya, Artemio Montesinos Palomino, contribution to the recognition of the publication: the translation into English and improvement in the writing.

Conflict of interest

The authors declare no conflict of interest.

Author details

Felipe Rafael Valle Díaz[1*], Oscar Apaza Apaza[2], Rosmel Iván Rodriguez Peceros[2], Alfredo Huamán Cuya[1], Percy Lima Román[1] and Artemio Montesinos Palomino[2]

1 University José María Arguedas, Perú

2 National University José María Arguedas, Peru

*Address all correspondence to: fvalle@unajma.edu.pe

References

[1] Peru. Regional government of Apurímac. Apurímac concerted regional development plan. In: Apurímac Regional Planning, Budget and Territorial Conditioning Management. 2021. Available from: https://app.regionapurimac.gob.pe/transparencia/wp-content/uploads/2023/01/PLAN%20DE%20DESARROLLO%20REGIONAL%20CONCERTADO%20-%20PDRC%20APUR%C3%8DMAC%20AL%202033.pdf

[2] Peru. Ministry of Foreign Trade and Tourism. Manual Updating and Inventory of Tourist Resources. Vice Ministry of Tourism. General Directorate of Tourism Strategy. Impresiones & Advertising Viserza SAC; 2019. Available from: https://cdn.www.gob.pe/uploads/document/file/401761/Manual_para_la_Elaboaracion_y_actualizacion_del_inventario_de_recursos_turisticos.pdf

[3] Peru. Official Gazette El Peruano. General Law of Peasant and Native Communities No. 24656. Congress of the Republic of Peru; 1987. Agrarian Affairs Commission, Official Newspaper El peruano. Available from: https://www2.congreso.gob.pe/sicr/cendocbib/con3_uibd.nsf/1DAB0BF2E43B8FBB0525797B006DE3C0/%24FILE/1_LEY_24656_Ley_General_Comunidades_Campesinas_SPIJ.pdf

[4] Peru. Official Gazette El Peruano. Law of Demarcation and Titling of Communal and Native Territories No. 24657. Congress of the Republic of Peru; 1987. Agrarian Affairs Commission, Official Newspaper El peruano. Available from: https://www2.congreso.gob.pe/sicr/cendocbib/con3_uibd.nsf/9BC059CD5D51F0A80525797B006E08FF/$FILE/3_LEY_24657_INTER%C3%89S_NACIONAL_COMUNIDADES_CAMPESINAS_SPIJ.pdf

[5] Valle Diaz FR. Potential of tourist resources as an opportunity for economic recovery in the Apurímac region, Peru, in the face of Covid-19. Ciencia Latina Multidisciplinary Scientific Journal. 2020;**4**(2):298-323. Available from: https://ciencialatina.org/index.php/cienciala/article/view/79

[6] Darcy S, Cameron B, Pegg S. Accessible tourism and sustainability: A discussion and case study. Journal of Sustainable Tourism. 2010;**18**(4):515-537. Available from: https://www.semanticscholar.org/paper/Accessible-tourism-and-sustainability%3A-a-discussion-Darcy-Cameron/7005bf6a64e7f0f7ab1e94fbe845563917e83ccf

[7] Molina H, Cánoves V. Accessible tourism, tourism for all: The situation in Catalonia and Spain. Tourism Notebooks. 2010;**25**:25-44. Mc Graw Hill/Interamericana Editores, SA de CV, Mexico. Available from: https://revistas.um.es/turismo/article/view/109561

[8] Fernandez Alles MT. Accessible Tourism: Analysis of Hotel Accessibility in the Province of Cadiz. [PhD thesis]. Cadiz University. 2007. Available from: https://rodin.uca.es/xmlui/handle/10498/15665

[9] Urquiola Sánchez O, Lalangui Ramírez J. The Banagua agro-tourism and nature route of the El Oro province. University and Society Magazine. 2016;**8**(3):128-134. Available from: http://rus.ucf.edu

[10] Miller A, Cox C, Morse. Ecotourism, agriculture, and wildlife conservation in Costa Rica through a social-ecologycal systems lens, 2. 2023. DOI: 10.3389/frsut.2023.1179887. Available from: https://www.frontiersin.org/articles/10.3389/frsut.2023.1179887/abstract

[11] Koshim A, Sergeyeva A, Kakimzhanov Y, Aktymbayeva A, Sakpbek M, Sapiyeva A. Sustainable development of ecotourism in "Altynemel" National Park, Kazakhstan: Assessment through the perception of residents. Sustainability. 2023;**15**(11): 2-18. DOI: 10.3390/su15118496. Available from: https://www.mdpi.com/2071-1050/15/11/8496

[12] Barragan J. Design of an ecotourism route in the municipality of Coloso - Sucre: Hiking and biodiversity [Master's thesis] Bolivar Technological University. 2017. Available from: https://biblioteca.utb.edu.co/notas/tesis/0070427.pdf

[13] Oseguera A, Fátima E, et al. Domestic group, territory and ecotourism in the community of Tziscao: Between tensions and conflicts. The Sustainable Journey. Oct 2022;(43):396-428. ISSN 1870-9036. DOI: 10.36677/elperiplo.v0i43.15288. Available from: https://rperiplo.uaemex.mx/article/view/15288 [Accessed: September 17, 2023]

[14] Hossa A, Darda A. Local perspectives of ecotourism development in the Terengganu state of Malaysia. 2023;**41**:1871-1200. DOI: 10.34624/rtd.v41i0.32418. Available from: https://scholar.google.es/scholar_url?url=https://proa.ua.pt/index.php/rtd/article/download/32418/22158&hl=es&sa=X&d=4449838490178367439&ei=g3p6ZOGmBIm5ygST0qGICA&scisig=AGlGAw9KL_ZjRqxKiTC733b9IAF6&oi=scholaralrt&hist=0DKWESIAAAAJ:5169405983547851934:AGlGAw-lKIqL1nhgqqMwXZi2PwLf&html=&pos=0&folt=rel&fols=

[15] Sidray B, Peña Y. Environmental Conditions for the Sustainable Development of Ecotourism in Santander. Colombia: Technological University of Santander; 2023. Available from: http://repositorio.uts.edu.co:8080/xmlui/handle/123456789/12491

[16] Castro Mirror. L. Proposal for an ecotourism route for the sustainable development of the Shitabamba hamlet in the Cutervo – Cajamarca National Park. [Professional degree thesis] Señor de Sipán University. Available from: https://repositorio.uss.edu.pe/handle/20.500.12802/4389

[17] Verdugo C, Montufar S, Andino E, Coronel M, Piray M, Andy A. Ecotourism as a cultural and sustainable business alternative in the Cavernas El Tocglo tourist center in Ecuador. Baltic Journal of Law and Politics. 2023;**33**(2): 1940-1958. Available from: https://www.namibian-studies.com/index.php/JNS/article/view/891

[18] Hernandez SR, Fernandez CC, Baptista LP. Research Methodology. Sixth ed. Mc Graw Hill/Interamericana Editores, SA de CV; 2014. Available from: https://www.uca.ac.cr/wp-content/uploads/2017/10/Investigacion.pdf

[19] Carrasco D. Scientific Research Methodology: Methodological Guidelines for Designing and Preparing the Research Project. 1st ed. Lima: San Marcos EIRLtda; 2005. Available from: http://www.sancristoballibros.com/libro/metodologia-de-la-investigacion-cientifica_45761

Chapter 10

Literature Review on South Asian Tourism Industry: Strengths and Weaknesses

Asma Akmal

Abstract

The purpose of this study is to evaluate the state of the tourism industry in South Asian nations. This study aims to examine the tourism potentials of South Asian countries, as well as the vulnerabilities and strengths of this industry. For analysis, the available literature was reviewed. According to the results, this region has the potential to attract both regional and international spiritual tourists. Researchers have identified a number of negative aspects, including inadequate transportation, infrastructure, service facilities, and untrained tourist personnel. To promote interregional tourism, it is necessary to revise visa and passport policies.

Keywords: spiritual tourism, thrilling tourism, economic growth, COVID-19, tourism weakness, strength

1. Introduction

South Asian nations include Bangladesh, India, Pakistan, Nepal, and Sri Lanka. These SAARC nations have common frontiers, cultures, and religions. Before 1947, the three significant countries Pakistan, India, and Bangladesh were united and known as the subcontinent. This region has been ruled by the Indus Valley Civilization, the Mughal, Sikh, and colonial empires. Every era has contributed a substantial number of historical and religious monuments, which now serve as tourist attractions. In 1947, the subcontinent was divided into India and Pakistan, and in 1971, East and West Pakistan were separated, giving rise to Bangladesh. These three nations, along with Sri Lanka and Nepal, are developing nations with similar economic, political, and social issues. The tourism industry in these regions had been neglected for years, and COVID-19 made the situation even worse. Travel restrictions and health and safety situations discourage both international and domestic tourism. The following statistics illustrate COVID-19's effect on South Asian nations (**Table 1**).

According to a report by the world bank, the graph above depicts the GDP and employment declines of South Asian nations during COVID-19. Efforts are made to discover methods for reducing these losses in the years following the COVID outbreak.

Country	GDP loss in COVID-19 US$	Job losses
Bangladesh	2.03 billion	420,000
India	43.4 billion	9 million
Nepal	460 million	230,000
Sri Lanka	1.99 billion	200,000
Pakistan	3.64 billion	880,000

Table 1.
Potential GDP and job losses in south Asian countries [1].

This chapter's objectives are:

- South Asia's tourism potential are examined.
- Evaluate the region's tourism industry's strengths and weaknesses.

These objectives will be accomplished by reviewing the relevant literature.

2. Literature review

A comprehensive literature review was conducted to evaluate the various characteristics of the tourism industry in South Asian nations. Following tourism categories were identified:

- Spiritual tourism
- Tourism revolving around thrill-seeking activities.

2.1 Investigations of spiritual tourism

The authors [2] wrote a book advocating for the promotion of spiritual tourism between India and Pakistan as a means of fostering regional harmony and economic development. They recommend that these two nations integrate spiritual tourism into their curricula, given the presence of a variety of religious minorities and their shared cultural and economic conditions. It is believed that by harnessing the force of spiritual tourism, terrorism can be mitigated and peace can be fostered in the region. In addition, the study [3] illuminates the significance of the Kartarpur corridor in attaining peace through tourism. It emphasizes the significance of cross-border encounters between visitors and their shared heritage sites in fostering regional harmony. Another study [4] offers a comprehensive overview of the Kartarpur corridor's development and economic contributions. It demonstrates that Pakistan is a popular tourist destination due to its renowned Sikh, Hindu, and Christian religious sites. In 2019, Pakistan's domestic tourism alone contributed substantially to the country's gross domestic product (GDP), which totaled $3.5 billion. This number was recorded before the Kartarpur corridor was opened. According to the research, the inauguration of the corridor will increase foreign exchange, trade, and employment opportunities. The study recommends the establishment of a federal department and provincial subdepartments to provide transportation facilities, promote local handicrafts, and increase religious tourism awareness. The authors suggest utilizing the tourism models of Nepal and the Philippines as benchmarks and guides. Another research study [5] conducted during the COVID-19 pandemic emphasizes the concerns of international tourists visiting India such as hygiene, safety, and security. These worries have caused substantial losses in the tourism industry. Therefore, the authors investigate the potential for religious sites to attract Indian domestic visitors. As a strategy for addressing crises in the tourism industry, the study suggests implementing Faulkner's (2001) crisis management model. It identifies the Swadesh Darshan and Spiritual Enhancement Drive (PRASAD) as tourism initiatives with the potential to attract domestic tourists. This innovative study offers valuable insights into tourism strategies post-COVID.

In addition, Pakistan has created a pilgrimage site for the Sikh community, and the government has constructed the Kartarpur route to facilitate the pilgrimage [6], used indicators such as perceived value (PV), sensation seeking (SS), low congestion (LC), as well as pilgrim satisfaction and behavioral intentions (BI), the study measures the level of satisfaction among these pilgrims. The findings indicate that PV, SS, and LC have a positive effect on BI and pilgrim satisfaction.

2.2 Investigations of adventure tourism

In an article [7], the author provides an in-depth analysis of Chitral Valley, an alluring tourist destination situated in the Hindu Kush Mountains near Trich Mir in Chitral, Pakistan. However, the Pakistani tourism department appears to be overly focused on the Northern regions, resulting in planning faults. These faults include insufficient seasonal security measures to mitigate seasonal disasters, limited foreign exchange benefits for the tourism sector, insufficient integration of local communities, insufficient funding, substandard service quality, and inadequate transportation facilities. Moreover, management flaws are evident, including an exclusive focus on the Northern region, poor infrastructure in terms of roads and railways, a lack of telecommunication facilities, high consumer goods prices, the absence of tourism courses in curricula, inadequate civic amenities, a lack of participation from local communities, and an absence of sports, beach, and cultural tourism initiatives.

An additional article [8], analyses briefly the dangers associated with tourism in the Himalayan mountains of India and Nepal. These regions attract visitors in search of hiking, sports, and adventure opportunities, but the climate poses a serious risk. On mountain slopes, additional dangers include rockfall, avalanches, and the flow of ice and/or snow. In these regions, roads are frequently eroded by severe weather conditions. Along with a rise in the number of commercial and sporting events, the number of tourists has increased dramatically over the years. This research indicates that the governments of these nations should maintain vigilance regarding the climate in these regions. In addition, tourist escorts and their families should be insured, and roads should be constructed in accordance with standards for mountainous terrain.

In a further article [9], the author concentrates on environmental management in tourism destinations, especially in developing communities such as South Asia. Rapid environmental changes are affecting these regions, necessitating the implementation of effective measures and policies by the governments of these nations in order to combat these changes and create sustainable environments. Due to political instability, inadequate institutional frameworks, and unequal social and economic structures, green tourism poses a significant challenge in this region. The paper suggests that the government, nongovernmental organizations, private sector policymakers, and the local community should collaborate to develop adaptable and simple strategies to promote green tourism.

2.3 Researches on tourism and economic growth

Javed and Tuckova [10] in an article presented at a conference examines the calculation of South Asian countries' competitive tourism advantage. In comparison to India and Bangladesh, the research indicates that Pakistan has emerged as a tourism destination, primarily due to its natural destinations and improved law and order situation over the past 4 years. Using the travel and tourism competitive index (TTCI) as an evaluation instrument, the study provides a comprehensive assessment of the region's tourism industry's strengths and weaknesses. It is suggested that Pakistan and India focus on environmental sustainability, Bangladesh prioritizes travel and tourism. Pakistan becomes more open to the international community, and Pakistan and Bangladesh focus on all aspects of infrastructure, while India focuses on service infrastructure.

Another analysis [11] focuses on the Bangladesh Tourism Board's (BTB) performance in obtaining a competitive edge in the region. While BTB policies have effectively increased domestic tourism over the past decade, attracting international tourists requires the translation of all marketing campaigns into English, the inclusion of hotels and restaurants in marketing initiatives, and the acquisition of international sponsorship. The BTB should re-evaluate its policies on a regular basis, improve the planning and management of tourist destinations, and identify barriers and obstacles to attracting international travelers.

In their research, the authors [12] note that the tourism volume of South Asian countries is significantly lower than that of other nations, accounting for only 1% of global tourism. This is primarily due to the lack of government involvement in providing infrastructure and facilities, which ultimately discourages private sector investment. The authors argue that the growth of the tourism industry is crucial to the prosperity of the people in this region. Using their shared borders, cultural heritage, and transportation resources, these nations can earn foreign currency, generate employment opportunities, and improve their economic well-being by promoting tourism.

Manandhar et al. [13] an additional comprehensive research describes the current state of the tourism industry in South Asian nations. Despite their shared borders, cultural heritage, and transportation resources, these nations' economies are hampered by geopolitical circumstances. Due to poor infrastructure, insufficient facilities and services, and safety and security concerns, including the threat of terrorism, the tourism industry in this region struggles to reach its full revenue potential. Complicated cross-border travel procedures, insufficient transport facilities, and a lack of coordinated travel policies are impeding the development of tourism in this region. The research suggests that these nations address these issues, revise their cross-border travel policies, and endeavor to maximize the economic benefits of the tourism industry.

During the COVID-19 pandemic, this study [14] provides a policy framework for policymakers in six South Asian nations. During the pandemic, tourism declined in these areas as it did in other nations. The study highlights the significance of information and communication technologies (ICT) in tourism growth. Due to the reluctance of international tourists to travel, the SAARC nations should waive visa requirements and offer discounts on travel and lodging costs for interregional tourists. During this pandemic, the importance of ICT in various facets of society, such as healthcare, education, and commerce, makes its active incorporation into the tourism industry crucial. With their export potential in fisheries, the research also suggests that Nepal and Bhutan should collaborate with regional and international commercial interests. In addition, another article by the same authors [15] supports the idea that, while tourism positively contributes to economic development, it can also have negative effects if not managed properly.

The authors [16] discuss the significance of private sector involvement in the tourism industry, as advocated by tourism scholars, especially in developing nations. The article [6] examines five policy options for South Asian nations, including public versus private participation, domestic versus international tourism, class versus mass tourism, centralized versus decentralized tourism, and integrated versus enclave tourism. Instead of rigorously adhering to the policies mentioned, these nations have taken a tourism-centric approach.

Munir and Iftikhar [17] studied the relationship between foreign direct investment (FDI) and exchange rate on tourism of five South Asian countries, India, Pakistan, Bangladesh, Nepal, and Sri Lanka. The role of FDI is proven from other researches also, and this research also validates that increase in FDI contracts tourism and decrease in it expands it in long run in all five countries in long run. Appreciation and depreciation in exchange rate have asymmetrical relationship with tourism. The analysis of individual countries showed that FDI has a negative and significant impact on tourism in India, Pakistan, and Bangladesh and has a positive and significant impact on tourism in Nepal and Sri Lanka. The exchange rate of all countries negatively effects tourism, increase in exchange rate decreases it and vice versa. FDI, according to this research, can increase tourism if it is spent on the improvement of tourism industry but unfortunately, these countries spend it on microeconomics challenges. It is highly recommended for these countries to spend FDI on the improvement of infrastructure, security and safety, and other facilities in tourism sector to gain economic prosperity.

Richter [18] describes the role of countries pollical conditions for the development of the tourism industry. The Philippines, Siri Lanka and Pakistan are facing problems of political instability and threats of war from neighboring countries. The study suggested risk analysis or political audit to devise appropriate tourism policies. It further commented that no matter how much tourism industry invests for new proposals, marketing, and services; if there will be political instability and threat of war, international tourist will not attract.

3. Discussion

This study's primary objective was to assess the tourism industry's potential in South Asia. A review of the literature revealed that South Asian countries are exceptionally abundant in art, architecture, and culture, which can attract tourists from around the globe. For instance, the Sikh community from India frequently traveled to Pakistan in search of spiritual fulfillment. Researchers concurred that these regions can generate foreign currency by promoting interregional/cross-border tourism. The researches of [2–4] supported the above description.

The second objective of this study was to assess the regional industry's strengths and deficiencies. This region's strong cultural, spiritual, and economic interregional connections make it a tourist-friendly destination. In Sri Lanka and Nepal [14], a variety of products, such as fish, have not only commercial value but also provide attractive fishing spots for tourists. Unfortunately, research identifies a lengthy list of vulnerabilities in this industry. First, the political situation [18] of these countries along COVID-19 is a significant barrier for international travelers. The law-and-order situation influences not only the decisions of international tourists but also those of interregional tourists. The countries are developing and have multiple microeconomic issues, so they invest little in the development of tourist destinations. Secondly, health and hygienic situations of these countries discourage international tourists. Interregional rivalries of these nations are third major issue in terms of biased visa policies for local tourists.

4. Conclusion

Literature review on tourism industry of south Asia indicated many important aspects of south Asian tourism industry. To understand characteristics of tourism industry this part is divided into two subheadings.

I. Strength of tourism industry

II. Weaknesses of tourism industry

4.1 Strength of tourism industry

The researchers agreed that these areas have large potential for tourism because of their beautiful sceneries, rich heritages, and culture. The strengths of the tourism industry are as follows:

- Spiritual tourism is one of the major sources of income for these areas as in the case of Kartarpur in Pakistan [4] and in different areas of India [5] and Bangladesh. The affluent class of these countries prefers spiritual journeys, just such as Muslims visit Mecca.

- These areas have great potential for cross-border tourism because of shared borders, resources, and cultures. In post-COVID-19 situations, these areas regional tourism can support their economy. So, they are not totally dependent on international tourists.

- Sri Lanka and Nepal are known for their fisheries industry. Fishing spots are attractive tourist points also [14].

- The currency rate of these areas are low in the international market so for international tourist, it will be economical, and for cross-border tourists, it will be affordable [17].

4.2 Weaknesses of tourism industry

The literature review identifies a number of weak points of South Asian Tourism Industry:

- This industry was neglected in the past [12] and became the center of focus specially in Pakistan in the last 4-5 years.
- All these countries need to revise their travel and tourism policies as mentioned by Mehraj Din Wani [6, 11] for Bangladesh T&T board.
- The infrastructure, safety and security, and hygiene of these areas are not up to international standards.
- The political situation in these countries, such as martial law and war-like conditions, is a major hindrance for international tourists [18].
- The complicated cross-border visa and passport procedures discourage regional tourists
- FDI (foreign direct investment) earned through tourism is spent on microeconomic issues rather than on the development of transport, infrastructure, service structure, and security of tourist points [17].
- Commercialization of hilly areas is creating problems of pollution and climatic safety [8].
- Climatic safety and security are a major issues for tourists in hilly areas [8].
- Information communication system is poor and unable to address target clients [14].

Author details

Asma Akmal
University of the Punjab, Lahore, Pakistan

*Address all correspondence to: asmaakmal4@gmail.com

References

[1] COVID-19 and Tourism in South Asia. Opportunities for Sustainable Regional Outcomes. World Bank. Available from: tourism@worldbankgroup.org. Social media. #tourism4dev

[2] Haq F, Medhekar A. Is Spiritual Tourism a Peace Vehicle for Social Transformation and Economic Prosperity in India and Pakistan? IGI Global Publisher of Timely Knowledge; 2019. DOI: 10.4018/978-1-5225-7464-4.ch010

[3] Salman Y. Tourism and reconciliation in an enduring rivalry: The case of Kartarpur Corridor on India–Pakistan border. Tourism Management Perspective. 2021;**39**

[4] Zaheer F, Mubariz S, Alvi ASAIF. Religious tourism backing for economic salvation: A. Journal of Indian Studies. 2020;**6**:127-136

[5] Manhas PS, Nair BB. Strategic role of religious tourism in recuperating the indian tourism sector Post-COVID-19. International Journal of Religious Tourism and Pilgrimage. 2020;**8**(7). Article 6

[6] Wani MD, Dada ZA, Shah SA. Behavioural intention and satisfaction levels among pilgrim tourists in South Asia. Tourism Review. 2023;**78**(4):1147-1163

[7] Bakhsh BQ. Managing tourism in Pakistan (a case study of Chitral valley). Journal of Managerial Sciences. **II**:169-190

[8] Ziegler AD, Wasson RJ, Sundriyal Y, Srivastava P, Sasges G, Ramchunder SJ, et al. A call for reducing tourism risk to environmental hazards in the Himalaya. Environmental Hazards. 2021:1-27

[9] Azam M, Sarker T. Green tourism in the context of climate change towards sustainable economic development in the south Asian region. Journal of Environmental Management and Tourism. 2011;**II**(1):6-81

[10] Javed M, Tuckova Z. Competitiveness in tourism: A comparative analysis of selected South-Asian countries. In: Portugal: Academic Conferences and Publishing International Limited Reading. Porto Portugal: 2nd International Conference on Tourism Research University Portucalense; 2019. pp. 381-388

[11] Kumar AS. An analysis of the activities and performance of Bangladesh tourism board (BTB). South Asian Journal of Social Studies and Economics. 2023;**17**:31-41

[12] Nawaz MA, Hassan S. Tourism in South Asia. International Journal of Economic Perspectives. 2016;**10**:591-601

[13] Manandhar, Rasul G, Prem. Prospects and problems in promoting tourism in south asia: A regional perspective. South Asia Economic Journal. 2009;**10**:187. DOI: 10.1177/139156140901000108

[14] Jayaraman TK, Makun K. Asymmetric analysis of tourism and economic growth in South Asian countries: Lessons for policymakers towards mitigating the adverse effects of Covid-19. Journal of Applied Economic Sciences. 2022;**XII**:49-55

[15] Keshmeer M, Jayaraman TK. Tourism and Economic Growth in South: Asymmetric Analysis and Lessons for Mitigating the Adverse. Munich Personal RePEc Archive; 2022. MPRA Paper NO. 111377, posted 06 Jan 2022 6:30 UTS.

Available from: https://mpra.ub.uni-muenchen.de/111377/

[16] Richter LK, Richter WL. Policy choices in South Asian tourism development. Annals of Tourism Research. 1985;**12**:201-217. DOI: 10.160-7383/85 $3

[17] Munir K, Iftikhar M. Asymmetric impact of FDI and exchange rate on tourism: Evidence from panel linear and nonlinear ARDL model. SAGE Open. 2021:1-10

[18] Richter LK. After Political Turmoil: The Lessons of Rebuilding Tourism in Three Asian Countries. Jtr. Sagepub.com at Pennsylvania State Univ.; 2016. pp. 41-45

Chapter 11

Tourism Education in the Post-Pandemic Era

Fangzhou Wan, Minghui Yang and Yubin Xi

Abstract

The field of tourism education has experienced significant transformations, shifting from its initial focus on hospitality management to encompass broader interdisciplinary perspectives. The COVID-19 pandemic has acted as a catalyst for transformative changes in higher education, leading to a rapid transition to online platforms and innovative pedagogical approaches within the realm of tourism education. Students have demonstrated adaptability in virtual learning environments, prompting institutions to reevaluate their pedagogy and management strategies. This crisis has underscored the importance of updating education to integrate both online and offline modalities, fostering resilience and versatility. In the post-pandemic era, the landscape of tourism education will witness profound changes in pedagogy, institutional practices, and student dynamics. Pedagogical innovations will involve integrating up-to-date industry knowledge, placing emphasis on practical skill development through experiential learning, and utilizing technology-driven channels for remote education. Institutional innovations will prioritize digitalization efforts along with health and safety measures while also focusing on sustainability and globalization initiatives. Educators will enhance their proficiency in utilizing new technologies as well as adopting interdisciplinary approaches. Student dynamics will emphasize international mobility opportunities, professional training programs, innovation mindset cultivation, and entrepreneurial skill development. Collaboration between academia and industry stakeholders will be crucial in preparing graduates for the evolving tourism landscape by emphasizing adaptability, continuous learning abilities, and forward-thinking mindsets.

Keywords: tourism education, pedagogical transformation, post-pandemic era, institutional innovation, interdisciplinary approaches

1. Introduction

The evolution of tourism education within higher education represents a significant paradigm shift [1]. Initially rooted in hospitality in Europe and predominantly focused on management and business perspectives in American programs [1], the field has expanded its scope over time. Noteworthy contributions from Asian countries such as Thailand, China, and Singapore have played a pivotal role in advancing tourism higher education since the previous century, highlighting the global significance of this academic domain [2].

However, the landscape of tourism education underwent a seismic shift with the onset of the COVID-19 pandemic, which disrupted higher education worldwide, including the realm of tourism education [3–6]. As physical gatherings became untenable due to health concerns and restrictions, educational institutions swiftly transitioned to online platforms, facilitating virtual connectivity between students and educators. This global crisis, while presenting myriad challenges, also catalyzed innovative pedagogical approaches, ushering in an era of ongoing evolution in tourism education [3–6].

Amid the pandemic, students demonstrated remarkable adaptability in acclimatizing to online learning environments, bolstered by widespread internet adoption globally [7]. This transition not only afforded students access to learning resources at a reduced cost compared to traditional tuition fees per credit hour but also compelled higher educational institutions to critically reassess their pedagogy and educational management strategies. Consequently, this introspection has ignited pertinent debates regarding the necessity of updating tourism education pedagogy to seamlessly integrate both online and offline course modalities, thereby fostering a more resilient and versatile educational framework.

The COVID-19 pandemic has profoundly reshaped various societal dimensions [8], with tourism education undergoing significant transformations [3]. As the world grapples with the aftermath of this crisis, it becomes imperative to scrutinize how education in the tourism sector will evolve in the post-pandemic landscape. Specifically, it prompts considerations on the nature of tourism pedagogy, the updating of educational management strategies by institutions, the adaptation of educators to evolving circumstances, the expectations and performance of students in the post-pandemic era, and the industry's expectations regarding graduates' competencies and skills. Acknowledging the significance of tourism education for both individual career development and the sustainable advancement of the tourism sector, this chapter elucidates the historical trajectory of tourism education, evaluates its pedagogical adaptations amidst the pandemic, delineates prospective pedagogical trajectories, and concludes by underscoring the pivotal role of adaptability and resilience in shaping the future trajectory of tourism education.

2. History of tourism education

Tourism, tracing its roots back to the nineteenth century with Thomas Cook's inaugural tour in 1841, has since blossomed into a global phenomenon, permeating both Western and Eastern societies [9]. Despite its longstanding presence in societal practices, tourism's academic inception is comparatively recent, characterized by its interdisciplinary nature and a dearth of a robust theoretical foundation [10]. Cornell University in the United States blazed the trail in 1927 by establishing the world's first undergraduate degree program in tourism education, subsequently inspiring the proliferation of certificate and degree programs worldwide [11].

In response to evolving educational demands and technological advancements, various educational institutions, ranging from vocational schools to comprehensive universities, have tailored programs catering to diverse levels of study within the hospitality and tourism domain [12]. The advent of online platforms has further democratized access to quality education, exemplified by initiatives like The Hong Kong Polytechnic University's MicroMaster program launched on the edX platform in 2017, which offers global participants a comprehensive hospitality and tourism curriculum [13].

As the 21st century unfolds, tourism education assumes a pivotal role within the broader landscape of social sciences. This evolution underscores the dynamic nature of tourism education, which has transitioned from its nascent stages characterized by a lack of theoretical grounding to its current status as a burgeoning academic discipline with a global footprint [10]. While some institutions integrate tourism education within their social science departments, others position it within business or management schools, underscoring its multifaceted nature and interdisciplinary significance [14]. As educational paradigms continue to evolve and adapt to societal needs, tourism education remains poised to play an integral role in shaping the future of global tourism and hospitality industries.

The evolution of tourism education has been influenced by socio-economic trends, technological advancements, and shifts in consumer behavior. Across the globe, institutions have recognized the need for specialized training in tourism management, hospitality, and related fields. Renowned institutions in Europe, such as the University of Surrey and the University of Nottingham, are engaged in developing research-oriented talents. Meanwhile, in Switzerland, institutions like École hôtelière de Lausanne Hospitality Business School and Les Roches Global Hospitality Management Education have set standards for excellence in practical hospitality education. Similarly, in the United States, Cornell University and the University of Nevada, Las Vegas, are renowned for their comprehensive programs in hospitality and tourism management.

Program category	Program purposes	Expected achievement	Duration	Future career orientation	Example institutions and programs
Training Program	Introduction to basic tourism concepts and skills	Basic understanding of the tourism industry	Short-term (weeks)	Entry-level positions in tourism and hospitality	TAFE Queensland – Tourism and Hospitality Skill Set
Certificate Program	Specialized skills and knowledge in a specific area	Specialized expertise	Several months	Specialized roles in tourism or related industries	Cornell University – Certificate in Hotel Revenue Management
Associate Degree Program	Foundational knowledge in tourism and hospitality	Fundamental understanding of tourism principles	2 years	Entry-level and mid-level roles in the industry	Santa Monica College – Associate in Science in Hospitality Management
Bachelor's Degree Program	In-depth study of tourism management and principles	Comprehensive understanding of tourism	3–4 years	Management roles in tourism, hospitality, and related sectors	University of Florida – Bachelor of Science in Tourism, Hospitality, and Event Management
Master's Degree Program	Advanced study in tourism, often with a specialization	Expertise in a specific area of tourism	1–2 years (after Bachelor's)	Leadership roles, academia, research, consulting	University of Surrey – MSc International Tourism Management
Doctoral Degree Program	Research-oriented, in-depth study and contribution to the field	Contribution to academic knowledge, research skills	3–5 years (after Master's)	Academic positions, high-level research, leadership in the industry	University of Otago – PhD in Tourism

Table 1.
General information of world's tourism education.

In Asia, educational hubs for tourism studies have emerged in countries like Singapore and Hong Kong, with institutions like the Hong Kong Polytechnic University offering specialized courses tailored to the region's growing tourism industry.

Table 1 provides a comprehensive overview of the categorization of tourism education programs globally, delineating each program category's objectives, anticipated outcomes, duration, career prospects, and exemplary institutions offering such programs. From short-term training initiatives introducing basic tourism concepts to doctoral programs fostering research contributions and leadership roles in academia and industry, the table underscores the diverse educational pathways available for individuals aspiring to pursue careers in the dynamic field of tourism.

3. Pedagogical transformation amidst the impact of the COVID-19 pandemic

The COVID-19 pandemic has undeniably exerted a profound influence on the tourism industry, necessitating unprecedented adaptations [3]. In response to travel restrictions and heightened safety concerns, tourism education institutions swiftly recognized the imperative for agile adjustments to equip students with the requisite competencies to navigate these exigencies [15, 16].

One significant response from these institutions has been the integration of technology into tourism education curricula [3, 4, 15, 16]. With prevalent travel restrictions and social distancing measures, virtual platforms and online tools emerged as pivotal in maintaining educational continuity. Students were exposed to innovative digital solutions, including virtual reality tours, augmented reality applications, and online simulations, facilitating remote exploration of destinations and skill practice.

Furthermore, the pandemic prompted an evolution in crisis management education within tourism programs, fostering a deeper comprehension of global health emergencies [17]. It served as a catalyst for the industry to prioritize public health considerations across all facets of tourism operations. Consequently, there emerged an augmented emphasis on topics such as epidemiology, infectious disease control measures, and emergency response protocols within tourism curricula.

In addition to crisis management and risk assessment, resilience training has emerged as a fundamental component of tourism education [18]. Students are now equipped with strategies to adeptly adapt to evolving circumstances and effectively rebound from disruptions. They receive instruction on contingency planning, business continuity strategies, and the development of robust recovery plans capable of withstanding future crises.

Moreover, there has been an amplified emphasis on sustainability within tourism education, spurred by the pandemic's profound impact on natural resources and local communities [19]. Students now exhibit heightened awareness regarding the significance of biodiversity preservation, reduction of carbon footprints, promotion of responsible consumption patterns, support for local economies through community-based initiatives, and engagement in ethical practices throughout their professional endeavors.

These adaptations underscore the resilience and responsiveness of tourism education in the face of unprecedented challenges. By embracing technological innovations, integrating crisis management education, fostering resilience, and prioritizing sustainability, tourism education institutions are not only preparing students to navigate the complexities of the present but also equipping them with the tools to shape a more resilient and sustainable future for the tourism industry.

4. Tourism education in the post-pandemic era

Following the conclusion of the pandemic, it is evident that the landscape of tourism education worldwide will undergo significant transformations. These changes will be characterized by emerging health and safety standards, technological advancements, and a renewed emphasis on sustainability and environmental concerns.

4.1 Pedagogical innovations

In the aftermath of the pandemic, there is an imminent expectation of a profound transformation in tourism education pedagogy. This transformation will be multifaceted, driven by shifts in teaching sources, methods, channels, students' expectations, industry demands, and educators' capabilities. Firstly, the sourcing of teaching materials will evolve to reflect the new realities of the tourism industry. Educators must incorporate updated information on health protocols, risk management strategies, and sustainable practices into their curriculum to equip students with the relevant knowledge and skills necessary for a resilient future in tourism. Secondly, a paradigm shift in teaching methods is anticipated. Traditional classroom lectures may be complemented or replaced by experiential learning approaches emphasizing practical skill development. Incorporating field trips or virtual simulations into tourism education programs can provide hands-on experiences and foster critical thinking among students. Moreover, technology-driven channels will play a pivotal role in delivering educational content, offering remote learning opportunities for students facing travel constraints or seeking flexible study options. Educators must adapt instructional techniques to effectively engage learners through digital mediums while ensuring accessibility for all. Post-pandemic, both students and the industry are likely to seek specialized knowledge in areas such as crisis management or sustainable tourism practices, alongside adaptable skills like data analysis or cultural sensitivity, reflecting evolving market demands.

4.2 Institutional innovation and educator development

The post-pandemic era will prioritize digitalization and technology-driven education, with continuous advancements expected to reshape tourism education. Technologies such as virtual reality (VR) and augmented reality (AR) are anticipated to provide immersive learning experiences, fostering interactive and engaging environments. Hybrid education, a blend of online and face-to-face learning, is poised to become a pivotal direction, offering flexibility and global access to courses. Higher education institutions may integrate adaptable pedagogical models to cater to students' needs for flexibility and interaction. Health and safety will become integral components of curricula, necessitating proficient implementation of hygiene measures and strategies for managing tourism operations to mitigate disease transmission risks. Additionally, sustainability will remain a focal point, driving the development and implementation of eco-friendly strategies in tourism education. Globalization and cross-cultural education will be strengthened, emphasizing effective communication and collaboration across diverse cultural backgrounds. Government policies may incentivize resource utilization to advance the industry. Consequently, educators must enhance their proficiency in utilizing new technologies and adapt to blended learning approaches. They should expand their knowledge base in health, safety, sustainability, and cross-cultural studies.

4.3 Student dynamics

Post-pandemic, a gradual recovery in international student mobility is anticipated, with countries adopting inclusive policies to encourage overseas education and cross-cultural exchanges. Tourism education will prioritize cultivating and attracting new talents through enhanced professional training and partnerships aligning educational content with industry demands. Fostering innovation and an entrepreneurial mindset will be crucial, achieved through relevant courses and projects aimed at creating novel offerings within the tourism industry. Future education may emphasize the development and management of local resources to deliver exceptional travel experiences for domestic tourists while promoting proximate tourist destinations effectively. In summary, future tourism education will emphasize health and safety measures, digitization, sustainability, globalization, government policy influence, human resource development, innovation, and entrepreneurial spirit. Pedagogical approaches should combine experiential learning with theoretical foundations, incorporating modules on crisis management, digital marketing, and data analytics. Partnerships between academia and industry stakeholders will facilitate internships and research collaborations, while mentorship programs provide exposure and networking opportunities. Embracing interdisciplinary perspectives will prepare graduates for the diverse nature of the global tourism industry while fostering cultural competency.

5. Conclusion

In conclusion, the COVID-19 pandemic has accelerated significant transformations in tourism education, prompting a reassessment of pedagogical approaches and curriculum development strategies. As the industry adapts to new realities and challenges, education must remain adaptable and responsive, equipping students with the knowledge, skills, and resilience required to thrive in an ever-evolving landscape. By embracing innovation, sustainability, and collaboration, tourism education can play a pivotal role in shaping a more resilient, inclusive, and sustainable future for the global tourism sector. As we navigate the uncertainties of the post-pandemic era, the importance of continuous learning, adaptability, and a forward-thinking mindset cannot be overstated in preparing the next generation of tourism professionals.

Future tourism education will place a greater emphasis on health and safety measures, the integration of digitization and technology, the implementation of sustainable practices in the tourism industry, the enhancement of globalization and cross-cultural communication skills, recognition of government policy influence on the industry's direction, adoption of human resource development strategies, and fostering innovation and an entrepreneurial spirit. Additionally, it will focus on developing and managing local and nearby tourism resources to cultivate adaptable and inventive talents capable of meeting the challenges and opportunities presented by the future of this industry. Looking ahead, tourism education must continue to evolve in response to emerging trends and global dynamics. Pedagogical approaches that combine experiential learning with theoretical foundations will enhance students' practical skills as well as their critical thinking abilities. Curricula should include modules on crisis management, digital marketing techniques, and data analytics to prepare students for post-pandemic realities in the tourism landscape. Furthermore, partnerships between academia and industry stakeholders will facilitate internship opportunities along with research collaborations, while mentorship programs provide students with invaluable exposure to the industry as well as networking prospects. Embracing interdisciplinary perspectives is crucial in preparing graduates for the diverse nature of the global tourism industry while also fostering cultural competency.

Conflict of interest

The authors declare no conflict of interest.

Author details

Fangzhou Wan*, Minghui Yang and Yubin Xi
College of Shanghai Lausanne Hospitality Management, Shanghai Business School, Shanghai, China

*Address all correspondence to: fangzhou.wan@connect.polyu.hk; fangzhouwan_angela@sbs.edu.cn

References

[1] Airey D. Tourism education: Past, present and future. Turističko Poslovanje. 2016;**17**:9-12. DOI: 10.5937/turpos1617009a

[2] Liu C, Schänzel H. Introduction to tourism education and Asia. In: Tourism Education and Asia. Perspectives on Asian Tourism. Singapore: Springer; 2018. pp. 3-11. DOI: 10.1007/978-981-13-2613-4_1

[3] Wan F, Hsu C. 'Love' or 'Hate'? TAM-guided Frameworks of Lecturers' and Students' Assessment of Online Teaching/Learning. Journal of China Tourism Research. 2022;**19**(3):517-544. DOI: 10.1080/19388160.2022.2096165

[4] Smith RA. Pandemic and post-pandemic digital pedagogy in hospitality education for generations Z, Alpha, and Beyond. Journal of Hospitality and Tourism Research. 2021;**45**(5):915-919. DOI: 10.1177/10963480211000818

[5] Leong CK, Yung SP, Hua AK, Chyi LS. Post-pandemic education: Hospitality program revisit. Journal of Teaching in Travel & Tourism. 2024;**24**(2):1-33. DOI: 10.1080/15313220.2023.2291755

[6] Xu J, Tavitiyaman P, Kim HJ, Lo S-KJ. Hospitality and tourism higher education in the post-COVID era: Is it time to change? Journal of Hospitality and Tourism Education. 2022;**34**(4):278-290. DOI: 10.1080/10963758.2022.2056044

[7] Guan J, Tavitiyaman P, Ren L, Wang P, Fang H. Hospitality students taking synchronous online classes: Are they ready? Journal of Quality Assurance in Hospitality & Tourism. 2022;**24**(2):121-145. DOI: 10.1080/1528008x.2021.2024779

[8] Hao F, Xiao Q, Chon K. COVID-19 and China's hotel industry: Impacts, a disaster management framework, and Post-Pandemic agenda. International Journal of Hospitality Management. 2020;**90**:102636. DOI: 10.1016/j.ijhm.2020.102636

[9] Tourism Service Suppliers. In Introducing the World's Largest Industry, Tourism. 2021. Available from: https://www.pearsonhighered.com/assets/samplechapter/0/1/3/4/0134484487.pdf

[10] Tribe J. The indiscipline of tourism. Annals of Tourism Research. 1997;**24**(3):638-657. DOI: 10.1016/s0160-7383(97)00020-0

[11] Nolan C. History. 2021. Available from: https://sha.cornell.edu/about/history/

[12] World Tourism Organization. UNWTO Tourism Education Guidelines. Madrid: World Tourism Organization (UNWTO); 2022. DOI: 10.18111/9789284423941

[13] MicroMasters in International Hospitality Management (Online Programme) | Taught Postgraduate Programmes | Study | School of Hotel and Tourism Management. School of Hotel and Tourism Management. Available from: https://www.polyu.edu.hk/shtm/study/taught-postgraduate-programmes_v2/micromasters-in-international-hospitality-management/

[14] Scott NM, Puleo VA, Crotts JC. An analysis of curriculum requirements among hospitality and tourism management programs in AACSB colleges of business in the United States. Journal of Teaching in Travel & Tourism. 2008;7(4):71-83. DOI: 10.1080/15313220802061014

[15] Edelheim JR. How should tourism education values be transformed after 2020? Tourism Geographies. 2020;**22**(3):547-554. DOI: 10.1080/14616688.2020.1760927

[16] Del Carmen Mínguez García M, Hernández CM, Bernabé C. Higher education and the sustainable tourism pedagogy: Are tourism students ready to lead change in the post pandemic era? Journal of Hospitality, Leisure, Sport & Tourism Education. 2021;**29**:100329. DOI: 10.1016/j.jhlste.2021.100329

[17] Liu X, Wen J, Kozak M, Jiang Y, Li Z. Negotiating interdisciplinary practice under the COVID-19 crisis: Opportunities and challenges for tourism research. Tourism Review (St Gallen). 2021;**77**(2):484-502. DOI: 10.1108/tr-01-2021-0034

[18] Prayag G. Tourism resilience in the 'new normal': Beyond jingle and jangle fallacies? Journal of Hospitality and Tourism Management. 2023;**54**:513-520. DOI: 10.1016/j.jhtm.2023.02.006

[19] Li H, Liu X, Zhou H, Li Z. Research progress and future agenda of COVID-19 in tourism and hospitality: A timely bibliometric review. International Journal of Contemporary Hospitality Management. 2022;**35**(7):2289-2321. DOI: 10.1108/ijchm-04-2022-0424

Chapter 12

Strategic Management Accounting Application – Some Discussions for the Tourism Business

Phan Thi Thuy Nga

Abstract

This chapter will introduce strategic management accounting (SMA), a modern analysis and forecasting tool that supports managers in making strategic management decisions to help businesses increase their performance in both financial and non-financial. By reviewing the literature and my empirical study about factors affecting SMA application in the tourism business, this chapter aims to help managers who work in the tourism business with some references and recommendations to decide on applying strategic management accounting, particularly SMA techniques. Following objectives of this chapter: - Understand strategic management accounting and strategic management accounting techniques - Understand the role of strategic management accounting in enterprises - Some propose applying SMA techniques in enterprises, specifically tourism businesses.

Keywords: strategic management accounting, SMA techniques, SMA application, performance, tourism

1. Introduction

The business environment is becoming increasingly fierce, requiring more information sources to compete and develop. Traditional management accounting gradually reveals many limitations in supporting managers because its information is only internal and exists in scope with a short-term forecast period. [1, 2]; emphasizes financial and historical information only [3]. Strategic management accounting was born as a necessity, supplementing and perfecting the limitations of traditional management accounting. Therefore, it is considered the era of analysis and forecasting because it helps businesses evaluate and forecast strategic visions to achieve their goals when the information it provides is more flexible with a broader scope beyond the "internal" of the enterprise [1, 4, 5].

Nearly 40 years since Simmond [3] first introduced strategic management accounting. There have been many debates, but the concept and content still do not have a unified framework. However, this tool is still encouraged to research and apply in many countries worldwide. If before the 2000s, related studies only tried to shape strategic management accounting, from 2000 to now, applied studies have become popular. The growing number of research studies was in developed economies such as the United States, United Kingdom, and New Zealand and developing countries such as Slovenia, Malaysia, Thailand, China, and Vietnam.

The correct awareness of the importance of SMA motivates managers to apply this tool more. For this application to become practical, many empirical studies also explore and test the analysis of influencing factors. Many factors are synthesized from previous studies, including characteristics of the CEO, CFO, the accountant's participation in the decision-making process of quality management, business strategy, decentralization, information quality of information systems, and technology;... Many models have been proposed to help managers evaluate their internal and external factors, thereby taking steps to apply this tool more effectively.

This chapter will briefly introduce conceptual perspectives on SMA, the primary content of each SMA technique will be presented based on a synthesis of related studies in the world through an empirical study in Vietnam. The author will present part of the recommended content to suggest the orientation of applying this tool in tourism businesses.

2. What is strategic management accounting?

Management accounting is an effective support tool for managers, helping them understand their business activities, strengths, and weaknesses to make short-term and long-term management decisions. When the environment becomes increasingly fierce, understanding the "internal" is an advantage to help businesses compete and allows businesses to interact with the outside. Simmond [3] realized that because he realized that profits are generated not only from manipulating information inside the company but also from external information, mainly from competitors. According to him, information related to competitors is essential because it helps businesses locate competitive advantages through cost comparison and cost structure collected from competitors. On that basis, he gave the first definition of the term Strategic Management Accounting (SMA), which is the provision and analysis of accounting data of competitors and enterprises to develop and monitor close to the competitive position [3].

However, SMA has not yet been clearly defined because, in traditional management accounting, it is still widely applied despite certain limitations in providing information to managers. Moreover, Simmond believes that the essence of SMA is accounting for strategic management, and the information to be collected by competitors must be financial and quantifiable. Therefore, he proposed to set up an accounting system to record data from competitors about costs, prices, and products as a basis for comparison to help businesses determine strategic positions.

Inheriting from Simmond but Bromwich [1] believes that SMA should focus on external issues, namely the need to collect more information from customers because he thought it was necessary to base on the product attributes that customers wanted, then produce products at the prices that customers will pay, helping to improve the competitive position of the business.

Simmond and Bromwich refer to SMA as content related to a specific audience, such as customers or competitors, but Lord [2] argues that SMA is a three-step process of collecting competitor information, exploiting opportunities to reduce costs, linking accounting to the strategic position and repeated in six phases; Dixon and Smith argue that SMA includes four stages: defining business strategy, strategic cost analysis, market analysis, and strategic evaluation [6].

This group of authors represents the first view of SMA when attempting to incorporate strategy into management accounting. That comes from the original meaning of the term "strategy" in the military, which implies understanding the opponent to win. Moreover, it was also the idea that, combined with Porter's strategy theory, flourished during that period. This view is once again agreed by the British Association of Certified Management Accountants when introducing the concept of SMA, which is the provision and analysis of management accounting data related to business strategy, especially relative and actual cost trends in terms of the firm's price, volume, market share, cash flow, and total resource needs.

Many definitions of SMA have been proposed in previous studies, summarized explicitly in the following **Table 1**.

In addition, the research literature review shows that the second view of SMA is a set of strategic-oriented strategic techniques to support enterprises to increase efficiency [4, 12], which is the concept used in most empirical studies. With the early form of SMA, the main techniques mentioned include cost-related plans and competitors [1, 4]. Roslender and Hart [13] extend the connection between management accounting and marketing management to enhance the marketing aspect of the concept of SMA. The results from this study show that in addition to cost-related techniques such as attribute costing, strategic costing, target costing, and activity-based costing, two standard methods and the balanced scorecard have been valued and connected with SMA to add value for marketing and sales as well as analysis of customer's ability to pay. Roslender and Hart [13] once again confirmed this point of

Author	Define
Simmonds [3]	SMA is the collection and processing of financial information of the management accountant on business activities, on the competitive situation of enterprises, and of the corporate investors to build a strategy and monitor the enterprise's strategy.
Shank [7]	SMA uses cost management information directed at each stage of the strategic management process.
Bromwich [1]	SMA is the analysis and provision of information on the product market, the cost structure of the enterprise, and the corporate investor to monitor the business quality of the enterprise and of the corporate investors in the same need for many periods (long-term).
Ward [8]	SMA is an accountant for strategic management.
Roslender and Hart [9]	SMA is a general approach to accounting for the positioning of business quality, defined by an effort to integrate insights from management accounting and marketing management within the quality management framework.
CIMA [10]	SMA is a management accounting system that provides information outside the enterprise, and the information source is created from within the enterprise to serve the process of strategic analysis, strategy selection, and strategy implementation.
Langfield-Smith [6]	SMA is the strategic direction for creating, interpreting, and analyzing management accounting information and corporate governance activities
Ma and Tayles [11]	SMA is a management accounting that provides a strategic-oriented information system for decision-making and control of strategic activities

Source: Collecting from previous studies.

Table 1.
Definitions of strategic management accounting.

view by analyzing each content constituting SMA in work "*Strategic management accounting: Lots in a Name?*". Despite the increase in the number of techniques, this view is opposed by some authors in the United States because they consider it to be just a loose association of methods in that set [14].

In general, SMA is understood as accounting for strategic management or management accounting to help enterprises implement business strategies and always has characteristics that show differences compared to traditional management accounting:

- SMA has an outward focus;
- Information of SMA is both financial information and non-financial information;
- the internal system or the set target system does not constrain SMA;
- Support orientation and strategy implementation for businesses;
- SMA is the direction to the future.

3. The role of strategic management accounting for business operations

Although there are many views, the role of strategic management accounting for businesses is indisputable because the value brought when applying this tool is always to increase the company's performance [12, 15, 16]. It is not only financial performance, such as revenue growth, profit, and return on investment, but also non-financial effects, such as customer satisfaction, product quality improvement, the development of market share, and new products [12]. Besides, it also helps businesses save and control costs. It even makes the information on the financial statements more relevant and timely, which most companies desire to be achieved when competing in a harsh environment.

In addition, the content and methods of each SMA technique have been completed, studied, and applied in practice quite a lot. Each technique, when used, helps businesses solve critical links that help managers collect, synthesize, process, analyze, evaluate, and forecast critical issues that companies need to achieve, such as costs, customers, suppliers, and strategies to achieve their primary goals. At the same time, applying SMA techniques will help managers have an internal and external overview to make more appropriate and effective management decisions. This problem is illustrated by the image (**Figure 1**) of the process of strategic management accounting with a broader coverage from the outside, from the environment, and always having interactions with each other instead of just focusing on each other. Focus on internal processes without paying attention to external issues like traditional management accounting. Moreover, as mentioned, the advantage of SMA is that the information collected and provided is both quantitative information, such as financial information, and qualitative data, such as non-financial information. That helps administrators effectively cover, analyze, handle, and forecast problems.

Therefore, management increasingly shows a general perspective and point of view, is more flexible and associated with leadership, helps increase value for customers and shareholders, and helps to support inspection, assessment, and resource

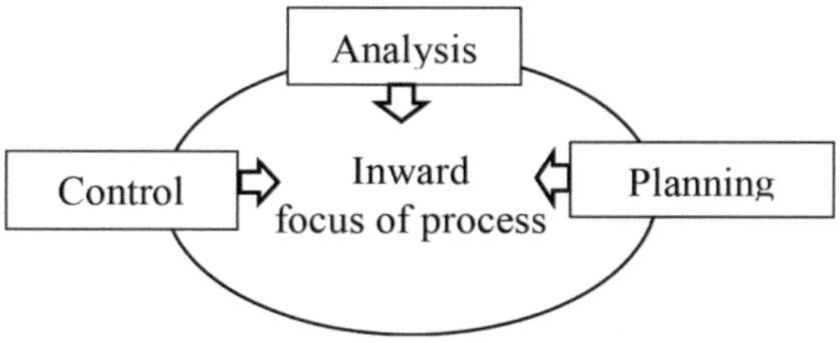

(a)

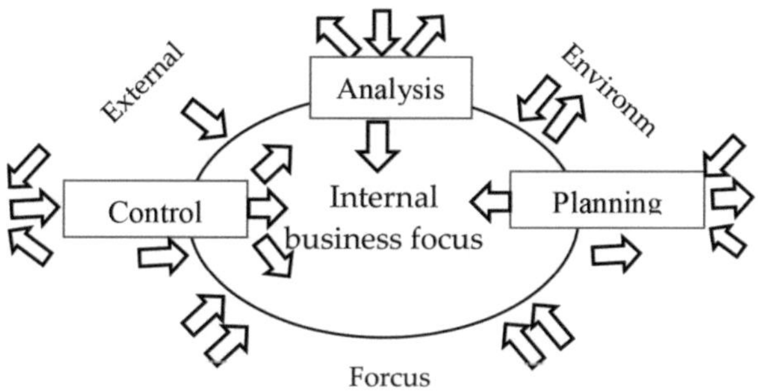

(b)

Figure 1.
(a) Traditional management accounting process and (b) strategic management accounting process. Source: Keith Ward [8], strategic management accounting, p. 87.

allocation. In particular, the role of SMA is more sophisticated in creating value through the efficient implementation of resources [17].

Thus, to make quality decisions, bring high efficiency, and help businesses achieve their goals, managers need financial and non-financial information collected not only in the past and present but also predictive of future problems. Moreover, it is collected from outside and inside enterprises to ensure two necessary and sufficient factors, quantity and quality factor of information when it is collected and provided.

4. Strategic management accounting techniques

With the role of providing information as a management tool, applying SMA and SMA techniques is a systematic process to record, process, synthesize and evaluate information related to specific objects and content. According to Cadez & Guilding [12], the SMA application has been widely recognized. From the quite popular results and the role of 2 new SMA techniques, in this study, the SMA techniques include 16 types according to Cadez & Guilding [12], adding two, namely ABC and environmental management accounting, bringing the total number of SMA techniques to 18.

Based on the concept of SMA introduced by the British Association of Notary Public Accountants [18], the application of SMA is the recording, processing, synthesizing, and evaluation of information related to 18 techniques according to the British Association of Certified Public Accountants [18] mentioned.

4.1 Costing

4.1.1 Attribute costing

This technique aims to collect, record and provide cost information showing the attributes and functions associated with the product in creating attraction and attraction for customers. Examples include the cost of value creation, the cost of using a warranty, the cost of helping to ensure delivery, and the after-sales service [1, 9, 12].

4.1.2 Life-cycle costing

This technique is collecting, recording, and providing cost information for each product life cycle stage, which is the total cost incurred related to each step as new design and introduction, growth, decline, and disappearance of a product/service to help businesses have a clear direction and strategy for the product [4, 12, 19].

4.1.3 Quality costing

This technique aims to collect, record and provide information about costs to produce products/services that meet specific quality technical standards businesses need to understand and control. Control costs include reasonable expenses (including fees of prevention and appraisal) and costs of non-conformities (costs incurred due to internal damage to the enterprise, costs incurred due to damage caused by internal damage to the enterprise, and external harm) help to allocate and use resources rationally and improve the quality of products/services [4, 12, 19, 20].

4.1.4 Target costing

This technique aims to collect, record and provide cost information to help businesses determine cost goals and the level of fees, cost savings, or appropriate use of costs. Based on the target selling price and target profit, the target costing process consists of three parts: market-oriented target cost, establishing and implementing the business quality, cost reduction, and target cost goals at the functional level [4, 12, 20].

4.1.5 Value chain costing

This technique is collecting, recording, and providing cost information identified in each activity of the value chain that creates value for the product: research, support costs & development, design, production, marketing, sales, distribution, and after-sales service. The value chain cost approach enables robust strategic planning, maximizing value while minimizing costs [12, 16].

4.1.6 Activity-based costing

It is a method of calculating costs based on activities that incur costs, thereby helping to trace the origin of activities that generate a lot of expenses and the relationship of costs with the object of charge. Costs to accurately determine the price and recognize the correlation and usefulness of the cost of the products and services businesses provide to the market [20].

DOI: http://dx.doi.org/10.5772/intechopen.112064

4.2 Planning, control, and performance measurement

4.2.1 Benchmarking

Standardization is a technique to set standards for costs, operations, and governance in the business and competitive market to help create conformity, improve costs, and operate. Business activities of enterprises are more suitable for corporate governance and show competitive advantages of specific enterprises of each activity, each division, and the whole enterprise [12, 16].

4.2.2 Integrated performance measurement (balanced scorecard)

A technique for measuring operational performance by connecting and concretizing vision and strategy into specific goals, measures, and actions on four dimensions. These include financial, customer, internal business processes, and the learning and growth team. This measurement technique helps to provide fully integrated financial and non-financial performance information, combining short-term with long-term, combining current with potential, and combining accounting with management [6, 12, 16].

4.3 Strategic decision-making

4.3.1 Strategic cost management

A system of public accounting associated with business quality to provide information to help enterprises gain competitive advantages based on comparing the costs of this enterprise with other enterprises. This technique is a process of identifying, accumulating, measuring, analyzing, evaluating, interpreting, and reporting cost information that is internally and externally relevant, group-related, and related to cost information. How organizations effectively use resources to improve competitive position and save enterprises costs [12, 16, 21].

4.3.2 Strategic pricing

This technique focuses on the factors that are likely to affect the price decision of the enterprise, such as corporate response, market share growth, price elasticity, customer value, and business environment, helping enterprises to react flexibly in pricing while ensuring competitive factors appropriate to the competitive position and implementing strategies of enterprises [12, 16].

4.3.3 Brand valuation

Including determining the budget and assessing the brand valuation to support management decisions in allocating resources, enhancing the brand position, and paying attention to the brand value. Pay attention to factors that affect the brand valuation, such as stability, leadership, market, international, trend, support, and protective factors related to brand profits history [12, 19].

4.4 Competition accounting

4.4.1 Competitor cost assessment

It is a regularly updated estimate of the costs and the composition of the costs of the public investment program in both quality and reliability to support the implementation of the cost reduction strategy, ensuring the prevention and control of costs that prevent fierce competition [12].

4.4.2 Competitor position monitoring

Analyze the position of the corporate market by collecting information about the corporate market's comparable sales, market share, volume, and cost. Based on that, enterprises can evaluate their competitive position compared to other companies from which to plan and control or build other business quality [12, 16, 21].

4.4.3 Competitor performance appraisal

this technique analyzes, synthesizes, and provides information published in the financial statements to the public, thereby evaluating their performance to make decisions [12, 16, 20, 21].

4.5 Customer accounting

4.5.1 Customer profitability analysis

Customer benefit analysis is the analysis of the cost of the service provided to the customer, including all activities to complete the sale and satisfy the customer, such as advertising, sales calls, deliveries, payments, collections, service calls, inquiries, and other forms of service to increase customer profitability and value that customer brings [12].

4.5.2 Lifetime customer profitability analysis

Similar to the above content, but this technique is associated with the classification according to the level of customer attachment to the business, based on which analysis of the ability to create value as well as the implementation cost for each type, this group of customers is more long-term and strategic [12, 16].

4.5.3 Valuation of customers as assets

Because customers are considered valuable and intangible assets of enterprises, this technique is associated with collecting information to evaluate, manage, and support the recognition of customers determining costs and the ability to generate profits from customers of enterprises [12].

4.6 Environmental management accounting

Environmental management accounting is a technique to provide information about costs related to the environment. It is a tool to support the implementation of environmental system construction to help establish an ecological information system for enterprises [22], simultaneously assessing benefits and costs and strategic planning for environmental governance (**Table 2**) [27].

5. Discussion

How to use SMA effectively? What factors affect the application of SMA? And what solutions do managers have to apply SMA effectively? To answer this problem, the author will rely on the results of my empirical study with the topic "Factors affecting the application of strategic management accounting - Research for tourism businesses in central Vietnam", this section will give some exchanges with readers, especially managers in tourism businesses about the application of SMA.

In the context of globalization, a sharing economy, and the current trend of linking and connecting stakeholders, the role and position of this tool are even more appreciated. To promote the value of adaptation into practice and increase the effectiveness of a strategic management tool in solving economic problems in the central provinces. The synthesized study tested the influence of factors on the application of SMA through a model established based on contigency and upper echelon theories. It was done by the mixed method of qualitative combined with a quantitative study. Data was collected from 326 tourism businesses in central Vietnam. Research results confirm the following factors: characteristics of the Board of Directors, Board of Directors; Decentralization, Competition, and Network Collaboration with differentiation strategies positively impact the application of SMA [26]. From the results of this study, several discussions are proposed for managers:

- Planning and developing the future management and management team from three angles:

The first is qualifications, and in addition to recruiting highly qualified human resources, enterprises also need to have a strategy to train human resources for each suitable position.

Secondly, as experience, enterprises need to assign many tasks in many positions and give opportunities to lead and manage groups when performing a task, project, etc., to have human resources to grasp the nature of operations.

Third, as leadership skill, this is an essential skill of a leader and manager of a business, a skill that will develop when trained through practical experience and the ability to learn from the people themselves. Therefore, enterprises need to recognize and create conditions for human resources to build those skills.

- Creating a competitive environment by assessing the level of competition to provide policies to support industry development and policies to support human resources, resources, and financial policies, regulate tourism development projects, etc., adjusting the competitive environment accordingly. In addition, when it comes to competitive advantage, it is also necessary to properly recognize competitive advantages and capabilities from the perspective of individual actions and cooperation to create benefits.

- Identifying issues that need to have cooperated with each object to improve the quality of cooperation; implement various forms of collaboration to increase

Strategic management accounting	Strategic management accounting techniques	Guilding et al. [4]	Lisa Jack [23]	Cadez & Guiding [12]	Cinquini & Tenncci [20]	Turner et al. [16]	Kalkhouran et al. [15]	Arunruangsirilert & Chonglerttham [24]	Hasid & Al1 – Sayed [21]	Marten et al. [25]	Phan Thi Thuy Nga [26]
		New Zealand and the United States	UK, Australia, Mỹ and New Zealand	Slovenia & Australia	Italia	Slovenia	Malaysia	Thailand	United Kingdom	Indonesia	Viet Nam
Costing	Attribute costing	×		×		×	×	×		×	×
	Life-cycle costing	×		×	×		×	×	×	×	×
	Quality costing	×		×	×		×	×	×	×	×
	Target costing	×		×	×		×	×	×	×	×
	Value-chain costing	×	×	×	×	×	×	×	×	×	×
	Activity-based cost				×		×	×	×		×
Planning control, and performance measurement	Benchmarking		×	×	×	×	×	×		×	×
	Integrated performance measurement		×	×	×		×	×	×	×	×
Strategic decision-making	Strategic costing	×		×		×	×	×	×	×	×
	Strategic pricing	×		×		×	×	×		×	×
	Brand valuation	×		×			×	×		×	×
Competitor accounting	Competitor cost assessment	×		×	×	×	×	×		×	×
	Competitive position monitoring	×		×	×		×	×	×	×	×
	Competitor performance appraisal	×		×	×		×	×	×	×	×

Strategic management accounting	Strategic management accounting techniques	Guilding et al. [4]	Lisa Jack [23]	Cadez & Guiding [12]	Cinquini & Tenncci [20]	Turner et al. [16]	Kalkhouran et al. [15]	Arunruangsirilert & Chonglerttham [24]	Hasid & Al1 – Sayed [21]	Marten et al. [25]	Phan Thi Thuy Nga [26]
		New Zealand and the United States	UK, Australia, Mỹ and New Zealand	Slovenia & Australia	Italia	Slovenia	Malaysia	Thailand	United Kingdom	Indonesia	Viet Nam
Customer Accounting	Customer profitability analysis	×		×	×	×	×	×		×	×
	Lifetime customer profitability analysis			×			×	×		×	×
	Valuation of customers as assets			×			×	×		×	×
Environment management accounting	Environment management accounting										×

Source: Collecting from previous studies.

Table 2.
SMA techniques application in empirical studies.

advantages and support each other's weaknesses to create quality products and services; have a vision and cooperation roadmap suitable to their conditions to improve service quality, and at the same time commit to, maintain and promote the value of cooperation; evaluate the quality of association to be transparent about benefits as well as eliminate if such collaboration does not bring any value.

- Agree on the division of responsibilities, assignment of tasks, and appropriate management decentralization to ensure the most effective operation of the management tool. Identifying and selecting business strategies helps managers effectively use available resources and synchronize resource use with the mission, vision, and plan to identify, analyze, implement, evaluate, and control effective strategy implementation and promotion.

- Understanding business characteristics to choose a strategy that is suitable for the conditions of the business as well as the current environment. From there, it is possible to establish a long-term strategy, have an adjustment roadmap, and ensure resources to implement them successfully.

6. Conclusion

Strategic management accounting is not a new concept, but it does not have a unified framework, methodology, or explicit content like traditional management accounting. Most studies applying strategic management accounting was based on the view that management accounting is a set of strategic management accounting techniques. However, theory and practice still have a laxity or phase difference.

The role and benefits of this tool have been verified in many studies worldwide. This chapter only introduces a basic concept of strategic management accounting and each SMA technique, which helps readers understand strategic management accounting and the role of SMA application in both theory and practice.

The recommendations for managers are based entirely on the results of my study on this topic. However, my study was for tourism businesses in Vietnam, so there will be different results from previous studies. Therefore, these recommendations only help readers to refer based on comparison to prior studies in the world to have orientations for researching or applying SMA in the future.

Author details

Phan Thi Thuy Nga
University of Finance and Accountancy, Quang Ngai Province, Vietnam

*Address all correspondence to: phanthithuynga@tckt.edu.vn

References

[1] Bromwich M. The case for strategic management accounting: The role of accounting information for strategy in competitive markets. Accounting, Organizations and Society. 1990;**15**(1): 27-46

[2] Lord. Strategic management accounting: The emperor's new clothes? Management Accounting Research. 1996;**7**:347-366

[3] Simmonds K. Strategic management accounting. Management Accounting. 1981;**5**(4):26-29

[4] Guilding C, Cravens KS, Tayles M. An international comparison of strategic management accounting practices. Management Accounting Research. 2000;**11**:113-135

[5] Kaplan RS. The evolution of management accounting. The Accounting Review. 1984;**59**(3):390-418

[6] Langfield-Smith K. Strategic management accounting: How far have we come in 25 years? Accounting, Auditing & Accountability Journal. 2008;**21**(2):204-228

[7] Shank JK. Strategic cost management: New wine, or just new bottles? Journal of Accounting Research. 1989;**12**(4): 487-506

[8] Ward K. Strategic Management Accounting. Oxford, United Kingdom: Butterworth-Heinemann, Published in association with the Chartered Institute of Management Accountants; 2004

[9] Roslender R, Hart SJ. In search of strategic management accounting: Theoretical and field study perspectives. Management Accounting Research. 2003;**14**:255-279

[10] CIMA. Chapter 8: The modern business environment. In: P2 Advanced Management Accounting - Study - Textbook. UK: Kaplan Publishing; 2019

[11] Ma Y, Tayles M. On the emergence of strategic management accounting: An institutional perspective. Accounting and Business Research. 2009;**39**(5): 473-495

[12] Cadez S, Guilding C. An exploratory investigation of an integrated contingency model of strategic management accounting. Accounting, Organizations and Society. 2008;**33**: 836-863

[13] Roslender R, Hart S. Strategic Management Accounting: Lots in a Name?, Accountancy Discussion Papers 1005. Accountancy Research Group, Heriot-Watt University (www.hw.ac.uk); 2010. pp. 1-27. Available from: https://www.researchgate.net/publication/266213810_Strategic_Management_Accounting_Lots_in_a_Name_Strategic_management_accounting_lots_in_a_name

[14] Nixon B, Burns J. The paradox of strategic management accounting. Management Accounting Research. 2012;**23**:229-244

[15] Kalkhouran et al. The indirect effect of strategic management accounting in the relationship between CEO characteristics and their networking activities, and company performance. Journal of Accounting & Organizational Change. 2017;**13**(4):471-491

[16] Turner M, Way S, Hodari D, Witteman W. Hotel property performance: The role of strategic management accounting. International

Journal of Hospitality Management. 2017;**63**:33-43

[17] Abdel-Kader M, Luther R. The impact of firm characteristics on management accounting practices: A UK-based empirical analysis. The British Accounting Review. 2008;**40**(1):2-27

[18] CIMA. Management Accounting in Support of the Strategic Management Process. 2005. Available from: www.cimaglobal.com.

[19] Cravens DW, Piercy NF. Relationship marketing and collaborative networks in service organizations. International Journal of Service Industry Management. 1994;**5** (5):39-53

[20] Cinquini L, Tenucci A. Strategic management accounting and business strategy: A loose coupling? Journal of Accounting & Organizational Change. 2010;**6**(2):228-259

[21] Hasid W, Al-Sayed M. Management accountants and strategic management accounting: The role of organizational culture and information systems. Management Accounting Research. 2021;**50**:100725

[22] Staniskis JK, Stasiskiene Z. Environment management accounting in Lithuania: Exploratory study of current practices, opportunities, and strategic intents. Journal of Cleaner Production. 2006;**74**(14):1252-1261

[23] Jack L. The adoption of strategic management accounting tools in agriculture post-subsidy reform: A comparative study of practices in the UK, the US. Australia and New Zealand, Chartered Institute of Management accountants. 2005;**5**(7):1-7

[24] Arunruangsirilert T, Chonglerttham S. Effect of corporate governance characteristics on strategic management accounting in Thailand. Asian Review of Accounting. 2017;**22**(1):85-105

[25] Marten PUYO, Sutrisno T, Rahman AF, Saraswati E. Environmental uncertainty: Typology of strategy and strategic management accounting techniques on company performance. International Journal of Environmental, Sustainability, and Social Science. 2022;**3** (2):489-503

[26] Nga PTT. Factors Affecting Strategic Management Accounting Application – An Empirical Study in Tourism Business in Central Vietnam [thesis]. Vietnam: University of Economic HCM city; 2023

[27] IFAC. International Guidelines on Environment Management Accounting (EMA). New York: International Federation of Accountants; 2005

Chapter 13

The Profile of Corporate Events Customers from the Organizers' Perspective in the Period Pre and Post-COVID-19 Pandemic

Cláudia Mendeiros and Rita Peres

Abstract

The present chapter aims to understand the current profile of the buyers of corporate events in the Cascais Municipality, from the organizers' perspective, to ascertain the pattern of customer demand and attitude toward two current trends: sustainability, as well as hybrid and online events, taking into consideration the evolution and change of the profile in the temporal context pre and post-2020 COVID-19 pandemic. After analyzing the results of the interviews, it is possible to verify that the profile of corporate clients is currently defined by a set of diversified characteristics, with a growing awareness toward sustainability and interest in hybrid events, as a complement to face-to-face events. It is also possible to highlight the impact that the COVID-19 pandemic and the consequent restrictions had in the evolution and change of the profile, such as the increased importance of flexibility in the conditions of contracted services, concern about sanitary measures and the demand for outdoor spaces.

Keywords: customer profile, corporate events, COVID-19, sustainability, hybrid and online

1. Introduction

There is a growing scholarly interest in events, indicative of their relevance in present-day tourism and their broader implications. This field of research is not only expanding in scope but also gaining a global perspective. There now exists literature focused on the impact of the events in cultural, social, and environmental aspects. However, there is a paucity of recent literature focused on the characterization and impacts of corporate events, particularly regarding the profile of the corporate client considering the various current trends. This study aims to develop this emerging field by exploring a better understanding of the buyers' profile of corporate events, regarding the trends of sustainability and hybrid and online events, in the period of pre- and post-COVID-19 pandemic.

"Corporate events, also known as business events, have multiple aims. Not only do they focus on creating added value to the businesses and individuals that are involved in their delivery (both participants and event providers), but also, on a macroeconomic level, they serve to support the growth and development of business tourism, thus boosting the local and national income, as well as the international profile of a particular destination [1]." This research is focused on the buyers, which in this case are conference organizers and meeting planners who buy, or more accurately, hire conference venues and related services to stage their events and who work for corporate organizations. Corporate organizations are companies established primarily to generate a profit, providing a financial return for their owners, whether they are the proprietors of a family-run business or the shareholders of a large publicly quoted company [2].

Considering that since the turn of the millennium, the trend of sustainability has been high on the agenda for many businesses, is visible the increased importance of corporate social responsibility (CSR) programs [3]. Many events' organizers attempt to incorporate an opportunity for the participants to participate in CSR-related activities and give something back to the communities they visit [3]. However, "research completed in the immediate prepandemic period (e.g. Dodds et al., 2020) suggests that event organizers already tended to ignore, or under-estimate, sustainability issues, and the pandemic may exacerbate such attitudes. A more positive forecast is that the crisis has encouraged people and organizations to appreciate the value of natural environments, livelihoods, and other people more, which may drive the adoption of more sustainable practices in the future [4]."

Regarding the online and virtual trend, "Seraphin (2020) and Madray (2020) have suggested that virtual events would be a solution for the event industry to cope with and recover from the COVID-19 pandemic. According to Barometer (2020), in terms of consumer behavior within the event industry, 84% of visitors and exhibitors have utilized one or more digital channels to use services since being forced to adhere to lockdowns and social distancing measures during the pandemic. 2020 saw an increase from 91% in June to 94% in September in the number of people willing to attend at least one event organized on any digital platform. PCMA Convene (2020) conducted a survey on business related to events professionals indicated that 87% of interviewees withdrew from their offline events; 66% of these events were delayed due to COVID-19, and 70% of companies switched their offline events to virtual events. Hence, digital events are projected to prevail over live or offline events [5]."

For these reasons, one of the main goals of this research is to understand the influence of the recent health pandemic COVID-19 on the behavior of corporate buyers related to sustainability and virtual trends, considering that, according to Madray [6], COVID-19 has irrupted the global markets and various industries drastically. However, innovative minds and leaders of respective fields struggled to bring new ways to survive in this condition. But still, the tourism and hospitality industries were severely affected, as the threat of disease has homed in the minds of the people and the policies of travel, entertainment, and events' organization and interlinked industries.

2. Global trends in corporate events

In recent years, the landscape of corporate events has undergone significant transformations driven by the ever-evolving global business environment. This chapter serves as an introduction to the subsequent exploration of the trends of sustainability and hybrid and online events in corporate events. By examining the dynamic nature of these trends, their impact on the industry, and the evolving preferences of corporate clients, including the influence of the COVID-19 pandemic, we aim to provide a comprehensive understanding of the current corporate events landscape.

Corporate events play a vital role in the modern business world, serving as platforms for networking, knowledge exchange, and brand promotion. The COVID-19 pandemic has brought unprecedent challenges and disruptions to the corporate events industry, reshaping the profile of corporate events and their expectations. According to American Express Meetings & Events [7], "Even though the events industry embraced virtual meetings during the COVID-19 pandemic out of necessity, they are not a substitute for face-to-face events. Face-to-face meetings are essential for collaboration, innovation, relationship building, and the development of trust needed to achieve common goals. Given that a considerable number of employees continue to work remotely, companies are challenged to keep their teams motivated and engaged."

"However, despite the current increase in the number of face-to-face meetings and events, virtual and hybrid formats are by no means disappearing. In fact, they are expected to play a key role in the future. Virtual technology is now seen as a valuable tool. Adding a virtual component to a face-to-face meeting allows for better management of attendee preferences, expanding event or meeting audiences, and even providing a contingency plan should it become necessary to change the format from face-to-face to virtual" [7].

In line with macro trends, organizations are increasingly considering sustainability issues as part of their base planning. There is a global consensus that sustainability should be a key factor in event programming [7]. By adopting eco-friendly strategies such as reducing waste, conserving energy, and incorporating social impact initiatives, corporate events can align with the values of attendees and stakeholders while leaving a lasting impression of environmental stewardship.

For these reasons, in the following subchapters the themes sustainability and hybrid and online events will be analyzed, framed in the universe of corporate events, to analyze the profile of the corporate customer in relation to these dimensions.

2.1 Corporate sustainability

In today's global context, where environmental concerns and social responsibility are at the forefront, sustainability has emerged as a key consideration in the planning and execution of corporate events, aligning with the global focus of not only minimize the ecological footprint but also contribute positively to the communities they engage with. Therefore, "corporate social responsibility is the voluntary dedication of companies to sustainable development, resulting in benefits to society (people), environment (planet) and economy (profit). At the consumer level, sustainability requires consideration not only of the price of a given good or service, but also of its impact on people and the natural environment" (UN, 1992; Jackson, 2005, cited in [3]).

Davidson [8] attributes part of the growth in interest in the social legacy of business events to the younger generation of participants, who are "uncomfortable with the type of 'conspicuous consumption' that can characterize such events when they feature bountiful, premier-winning and dining and lavish entertainment – often against a background of the deprivation of disadvantaged communities or in developing countries." According to the author, heightened social awareness, particularly among younger, "Generation Y" participants, and their desire to somehow "make a difference" and "give something back" to the communities where their corporate meetings and incentive trips take place has been a key factor in the growth of social legacy initiatives accompanying business events, and it is one of the key trends driving the design of such events today. The author links this phenomenon to the growing interest in volunteerism (or "volun-tourism") as a form of community service that promotes goodwill and is also personally fulfilling, adding that, for many participants, the social legacy element has also become an essential part of the "experiential" dimension of participating in a meeting or incentive trip.

In this sense, events can also aim to educate and influence their audience to adopt more environmentally conscious behavior after the event, to promote long-term benefits for the community in question [9]. Moreover, sustainability in corporate events showcases an organization's commitment to long-term sustainability goals, enhances its reputation, and positions it as a leader in responsible business practices.

In the context of corporate events, there was an important development in 1997: the adoption of Elkington's concept of the Triple Bottom Line (TBL) within the events sector. This is a tool that aims to highlight the areas of impact that a particular event can have. The tool is useful not only in the planning stages of an event but also at the event evaluation stage to identify what the impacts were. John Elkington's Triple Bottom Line proposal was embraced by several organizations that, through this tool, created corporate measures and reports on their economic, environmental, and social impact – or people, planet, and profit [3].

Merrilees and Marles (2011, cited in [8]) point out that given the significant footprint that corporate events can leave on the natural environment, this industry has increasingly responded to pressure to demonstrate its sustainable credentials. However, despite the magnitude of this issue, the amount of academic research that has focused specifically on sustainability in the corporate events industry has been minor, particularly when compared to the extensive discussions related to sustainability and tourism in academic publications. The explanation may be those commercial and financial aspects dominate in this area, so environmental considerations take a back seat.

However, Davidson [8] notes that there have been some notable contributions to the literature on this topic in the corporate event meetings sector: Park and Boo (2010) conducted a study on the green attitudes associated with two US conventions, primarily from the perspective of attendees, but also from the perspective of convention organizers and suppliers. Other valuable insights came from research by Draper *et al.* (2011), Mair and Jago (2010), and Mair (2011), who developed a model for greening the corporate sector. Their model is governed by organizational goals and values in an economic context, defining benefits and obligations through regulations and ranking the organization with different levels of green practices.

Thus, it is concluded that sustainable events are those that fulfill important social, cultural, economic, and environmental roles, with a positive impact that extends to the community of the place where they take place (Getz, 2007, cited in [3]). Embracing sustainability in corporate events is not only a strategic decision but also a moral imperative, as it reflects a collective effort to create a more sustainable future for generations to come.

2.2 Hybrid and online events

The recent importance of hybrid and online events in the corporate events landscape cannot be ignored. With the advent of technology and the challenges posed by the COVID-19 pandemic, the shift toward virtual platforms has become a necessity for the industry. Hybrid and online events offer several advantages, including expanded reach, cost-effectiveness, and flexibility. They allow businesses to connect with a global audience, breaking down geographical barriers and enabling participation from remote locations.

Consumer experiences in virtual environments are a preponderant factor for the success of the consumer market in several sectors, such as the organization of events. This concept has been widely explored in marketing (Klaus & Maklan, 2011, cited in [5]). According to Getz (2008, p. 404, cited in [5]), the main attraction of events is that each event is inimitable, and one must be present to fully live this unique experience. Communication and experiences are crucial aspects in virtual events because they can add interest and value to participants and to the hospitality and tourism industries. Moreover, creating positive online experiences for participants is essential to the success of an event (McGillivray, 2014, cited in [5]). According to Kharouf *et al.* (2020, cited in [5]) a good online experience can produce positive behavioral intentions toward the event. The authors defined the three factors with the most impact on consumers' online experiences: the interactions between participants, effective communication, and the level of online engagement. Moreover, this format provides opportunities for data collection and the ability to measure event impact and return on investment more effectively.

As the world of virtual and hybrid meetings continues to gain popularity within hospitality and tourism, there is an immediate need for additional literature in this area of study. Researchers could begin to look at reality outside of the hospitality and tourism industry, using an interdisciplinary approach to deepen knowledge within this area, but ultimately, a steady stream of research must be developed and extended in hospitality and tourism studies (Kim & Park, 2009, cited in [10]). Fryatt *et al.* (2012, cited in [10]) reinforce this idea considering that because virtual meetings are rapidly becoming more common and hybrid meetings are being touted as the future of meetings and events, it is imperative that academic research keeps pace with these developments.

While virtual events experienced a surge during the peak of the Covid-19 pandemic due to restrictions on in-person gatherings, there has been a shift toward the return of in-person components in the corporate events scenery. As the American Express Meetings & Events international study [7] shows that face-to-face events reiterate their importance in the 2022 forecast, considering that the forecast suggests an increase in the intention to hold events in this format.

However, despite the clear decrease in virtual events in 2022, hybrid events maintain the same level of interest, which suggests the consistency of their trend in the future. This suggests that the hybrid format offers the best of both worlds, catering to the diverse needs of attendees, whether they prefer to participate in person or remotely.

According to Nilsson [11], there are several ways to hold meetings and events online; however, it is important to analyze which format is best depending on the purpose of the meeting or event. Face-to-face meetings are characterized by the fact that participants know each other personally, and the factors inherent to this particularity should be considered.

When Fryatt et al. (2012) did their research, they found that some certain events and meetings were perceived to be better suited for the face-to-face or virtual environment. It was found that internal meetings could be held online, and, on the other hand, meetings about more sensitive topics, incentives, and celebrations would be more successful if held in person (cited in [11]).

Hybrid and online events have proven to be invaluable tools for corporate events, providing a dynamic and accessible platform for knowledge sharing, networking, and brand promotion in an ever-evolving digital panorama. The integration of both online and offline components ensures that organizations can leverage the advantages of both formats to deliver impactful and inclusive experiences.

2.2.1 Online events

Continuing the topic of online events, they have played a crucial role in the corporate events landscape, especially during the height of the COVID-19 pandemic. While there has been a recent decrease in virtual events as restrictions ease, it is important to recognize the lasting impact and value that online events have brought to the industry.

A virtual meeting is defined by Rubinger *et al.* (2020, cited in [5]) as an event that is meticulously planned by event organizers and attended remotely by participants, unlike a conventional physical event. Meanwhile, in the chapter "What is a virtual event?" (2020, cited in [5]), it is mentioned that virtual events can mimic physical events through online and virtual platforms, thus allowing an individual or an organization to hold various types of events (such as international events, conferences, trade shows, product launches or recruitment fairs) from their home or office, using a laptop or even a smartphone with an internet connection.

Additionally, Howard (2020, cited in [5]) mentions that considering that events may be composed of a single meeting or a set of several meetings, virtual events may complement the main event, thus not being a complete replacement of physical events.

According to Nilsson [11], there are eight different formats at the technology level for virtual meetings: videoconference, teleconference, online meeting, webcast, webinar, virtual fair, 3D virtual world, or hybrid meeting. The choice of format must be analyzed to define which technology is most suitable for the purpose of the meeting or event in question.

Online events have proven to be a lifeline for businesses, allowing them to maintain continuity and adapt to the new normal of remote work and travel limitations. They have provided a means for organizations to connect with their target audience, engage with participants across different time zones, and expand their reach globally. Online events have also demonstrated their cost-effectiveness, as they eliminate expenses associated with travel, accommodation, and venue rentals. Additionally, the digital nature of online events enables data collection and analysis, providing valuable insights into attendee behavior, preferences, and engagement levels. While the shift toward hybrid events signifies the importance of in-person interactions, the significance of online events should not be underestimated, as they have restructured the way corporate events are conceived and executed.

2.2.2 Hybrid events

Expanding on the topic of hybrid events, their continued relevance and interest in the corporate events industry can be attributed to their ability to bridge the gap between in-person and virtual experiences. The capacity to reach a wider audience through virtual components while still providing the benefits of physical presence positions hybrid events as a powerful and adaptable solution. This consistency in interest highlights the significance of hybrid events as a long-term trend, demonstrating their potential to shape the future of corporate events by providing a flexible and inclusive experience for all participants.

As Nilsson [11] mentions, the format creates an opportunity to add value by reaching a greater number of participants (in person and virtually) and by providing new formats of content and communication. The technology used in hybrid events allows the sharing of content, ideas, and experiences with participants in various regions and different time zones around the world, although several people point out negative points such as technological flaws and potential cannibalization of face-to-face events. The main advantage of hybrid events is the greater reach of the message to be transmitted and the redirection of part of the content, so that everyone can participate, including those who cannot travel to the venue.

This enhanced accessibility, combined with the unique opportunities for networking, knowledge sharing, and immersive experiences offered by in-person interactions, positions hybrid events as a compelling choice for both event organizers and participants. Fryatt *et al.* (2012, cited in [11]) state that hybrid meetings integrate technology with traditional event practices to create new types of experiences for participants and new ways of exposing content. The hybrid format is thus characterized by any meeting or event with at least one face-to-face participant group that connects digitally with participants located in another or various other locations around the world. The combination of the characteristics of face-to-face events and virtual meetings in hybrid events allows event organizers to achieve a greater breadth in terms of forms of participant participation.

We thus conclude that hybrid events are expected to maintain their significance and influence by offering a blend of physical and virtual experiences that cater to the evolving needs and preferences of attendees.

3. Material and methods

Considering the aim of the investigation, the methodological approach was divided into four main phases ending in the assessment of the corporate events customer profile toward the trends of sustainability and hybrid and online events, in the context of pre and post-COVID-19 pandemic (**Figure 1**).

In this research the qualitative method was selected. The qualitative method focuses on interpretation rather than quantification. Therefore, the researcher is more interested in understanding the interviewees' interpretation of the topic. Consequently, it is a method that emphasizes the subjective over the objective [12].

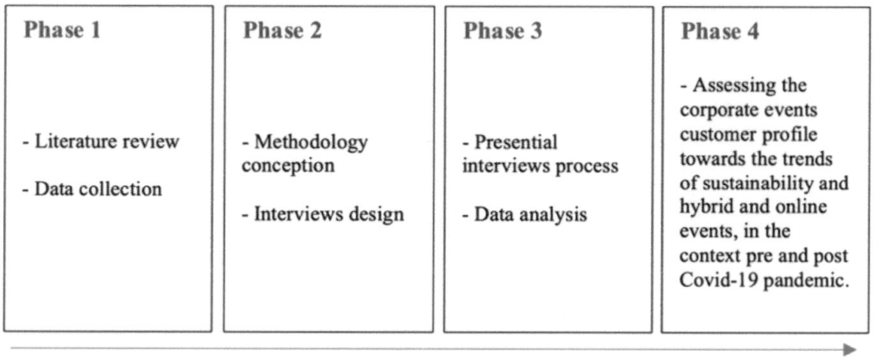

Figure 1.
Methodological approach.

Additionally, qualitative research pays attention to context, "in the sense that people's behavior and the situation link closely in shaping experience" [13].

The interview was structured according to three groups of questions about sustainability, hybrid, and online events, as well as the evolution and change of the corporate events customer profile in the pre and post-COVID-19 pandemic context, to develop these three perspectives of the study.

Over a period of two months (from May until July 2022), 11 structured interviews were conducted with organizers or those responsible for the events department of various companies or entities.

Researchers must have contact with the context in which they intend to conduct their research [14]. For this reason, interviews were conducted with corporate event organizers in the Municipality of Cascais, with direct contact with customers in this segment of events.

The interviewees were selected for their broad view of the process of organization and preparation of the event, to provide the necessary data for the study of the established objectives. Being from different types of companies, involved in the organization of corporate events (from event spaces, such as hotels, and event companies), they can thus provide different viewpoints on this research (**Figure 2**).

The sample for this study instrument consists of 11 interviewees, as shown in the graph above, whereby:

- Five interviewees are from hotel units located in the Municipality of Cascais, which host and organize corporate events in their spaces.
- Two respondents are associated with venues, in the Municipality of Cascais, which host and organize corporate events there (one of whom is an employee of a hotel unit that provides the service of organizing events there).

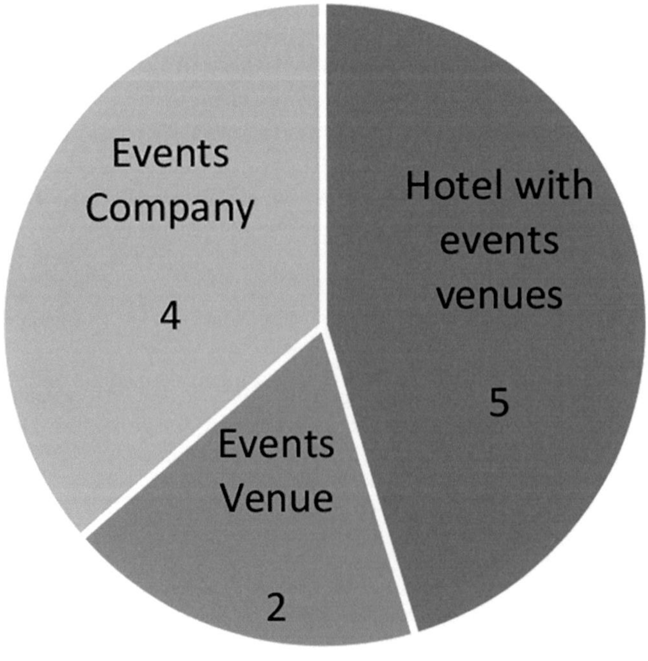

Figure 2.
Origin of the interviewees.

- Four interviewees are from events companies that organize corporate events in various venues in the Municipality of Cascais (one of whom works in an events company that is also a DMC, and another interviewee is an employee in an events company specializing in corporate retreats).

According to Vergara [15], the sample is a part of the universe chosen according to some criterion of representativeness. Thus, the sample refers to a subset of the population that is representative of the main areas of interest of the study [16]. Thus, the creation of this methodological tool was intended to acquire the indispensable information for the preparation of this research, given the reality of the market and its players.

3.1 Study area: Municipality of Cascais

The Municipality of Cascais is inserted in the Lisbon tourism region, located on the Estoril Coast. The Municipality of Cascais was chosen as the analysis area for this study, as it is a very expressive area in terms of tourism supply and demand, considering that already "in 2007, Estoril was the fourth national tourist destination", "distinguished by the high number of four- and five-star units" [17].

According to Casimiro *et al.* [17], "the promotion of the Estoril brand focuses on its main products (Leisure, MICE, and Golf)." Being the corporate events industry, included in the MICE segment.

In a recent interview with the president of Cascais Tourism, it is mentioned that "business travel represents only 21% of global travel, this refers to 2019. But it is important to point out that it is also one of the types of travel that spends the most in many of the destinations. Therefore, it is essential (...) for the whole tourism industry" [18].

Barros [18] also adds that "regarding business tourism, it is naturally a great asset and has the advantage of not being seasonal. And Cascais has all the conditions to be a reference destination in terms of business tourism. For these venues that receive them, (...) an investment in the infrastructure of the venues themselves is foreseen, which leads to the attraction of more and better events and meetings and, consequently, more business visitors to Cascais. We must always bear in mind that the business visitor may be our tourist of tomorrow."

According to a Visit Cascais [19] survey, carried out during the high season from April to September 2019, the motivation of travel by business represents around 14% of respondents, thus occupying second place.

Considering the data presented and the fact that business tourism is one of the main products of the Municipality of Cascais, it should be noted that this study aims to add knowledge regarding this segment, which not only has a significant current expression, but is also expanding, in the geographical area in question.

4. Interviews and results

This chapter will analyze the results obtained from the 11 interviews conducted with event organizers who will give their point of view on demand regarding the profile of the corporate events customer in the Municipality of Cascais.

The interview was essentially divided into the following topics regarding the profile of the corporate events client:

1. Their behavior and interest profile in the face of two current global trends:

   a. Sustainability.

   a. Hybrid and Online Events.

2. The evolution and change of the profile in the pre and post-2020 timeframe (COVID-19 pandemic).

4.1 Sustainability

As per **Table 1**, from the interviewees' perspective, it was possible to understand that a reduced percentage of customers (until 25%) show a preference for a destination and venue with sustainable practices (questions A and B). The same reduced percentage of 25% is applicable for the percentage of customers with interest in social legacy and corporate social responsibility activities (question C).

Concerning the yes or no questions, it was possible to identify that most of the customers are unwilling to pay more for sustainable services, even though four interviewees mentioned the opposite (question D).

It was also possible to ascertain that most of the corporate clients do not show considerable knowledge about sustainability certifications (question E), nor make it known if the company has sustainable policies (question F).

	Questions*	Questions with Percentages					Do not know or not applicable
		0%	1–25%	26–50%	51–75%	76–100%	
Quantity	a	1	7	0	2	0	1
	b	1	**6**	0	3	0	1
	c	0	**6**	2	1	0	2

	Questions**	Yes or No Questions		
		Yes	No	Do not know or Not applicable
Quantity	d	4	**6**	1
	e	2	**9**	0
	f	4	7	0

The highest values found are in bold.
(a) What is the approximate percentage of customers who show a preference for a destination with sustainable practices? (b) What is the approximate percentage of clients who show preference in hiring a venue (hotel, congress centre, etc.) with a sustainable practices certificate? (c) Regarding customers with an interest in sustainable practices, which percentage is interested in "Social Legacy" and "corporate social responsibility" actions and activities?
**(d) In the chosen space, do customers show a willingness to pay more for 'sustainable' services? (e) Regarding clients with an interest in sustainable practices, do they show any knowledge regarding sustainability certifications in the events industry? (f) Do they make it known if the company itself has sustainability policies?*

Table 1.
Closed-ended questions about sustainability.

4.2 Hybrid and online events

As per **Table 2**, from the interviewees' perspective, it was possible to verify that there was a very high interest (between 76% and 100%) in the online and hybrid events since COVID-19 (question A) and that nowadays, most of the events take place in a face-to-face format (question B). Concerning the online and hybrid formats, we can ascertain that they occur less frequently than the face-to-face events, even though the hybrid format has a stronger presence compared with the exclusive online events.

Two open questions about the hybrid and online events were also part of the interview. First, we asked the organizers to let us know the main types of events when organizing hybrid and online events. The internal meetings type was mentioned a total of ten times by the interviewees, and we could therefore consider it the category of event that most occur in online and hybrid formats. Then we asked which types of events continue to be face-to-face due to a lack of interest in hosting them in online and hybrid formats. The most frequent types of events mentioned were the incentives (six times), the team buildings (five times), and the celebrations (four times); therefore, we could consider that the corporate buyers prefer the face-to-face format for these event types.

4.3 Pre and post-COVID-19 period

Regarding the pre and post-COVID-19 period, was formulated an open question asking to the interviews to describe the main change or changes regarding the profile of the corporate customer in the pre and post-COVID-19 pandemic, within the scope of sustainability and hybrid and online events.

The changes most frequently mentioned by interviewees were the last-minute event requests, with a considerably shorter preparation time (mentioned by eight interviewees), and the greater attention to cancelation and reduction contractual clauses, with more flexible conditions (referred to by seven interviewees).

Additionally, several interviewees also referred the increased concern about health measures regarding COVID-19 (referred to by five interviewees) and the higher demand for spaces close to nature and with less tourist pressure (mentioned by three interviewees).

Another aspect that was mentioned by some of the interviewees was the current importance of technology, considering the adherence to the hybrid format as a complement to face-to-face events.

Questions*			**Questions with Percentages**				
			0%	**1–25%**	**26–50%**	**51–75%**	**76–100%**
Quantity	a					1	**10**
	b	*Face-to-face*	0	1	0	1	**9**
		Online	1	**10**	0	0	0
		Hybrid	1	**6**	1	1	2

The highest values found are in bold.
**(a) What is the approximate percentage increase in the level of customer interest in online and hybrid events since COVID-19? (b) What is the approximate percentage of face-to-face, online and hybrid events currently?*

Table 2.
Closed-ended questions about hybrid and online events.

5. Discussions, conclusions, and practical implications

After analyzing the results obtained through the interviews carried out with corporate event organizers, we could identify that the preference for a destination and spaces with sustainable practices is not yet demonstrated in a relevant way by the corporate buyers, despite being a growing concern and trend, and most corporate clients do not have in-depth knowledge about sustainability certifications. Therefore, we can conclude that, even though the corporate industry has increasingly responded to pressure to demonstrate its sustainable credentials [8], the absence of sustainable practices and credentials do not assume a deal breaker meaning for the corporate event buyers, even though there is a growing awareness and preference for sustainable procedures.

It was also possible to verify that there is interest in "Social Legacy" and "corporate social responsibility" actions and activities by corporate clients but still with little relevance. One of the interviewees mentioned that corporate buyers usually have a preference for social legacy activities when they are suggested to them; however, if not presented, many customers do not ask about this type of activity. This proves their desire to somehow "make a difference" and "give something back" to the communities where their corporate meetings and incentive trips take place [8], but still not in a very relevant way.

In this line of thought, it was also demonstrated that only a reduced percentage of corporate clients is willing to pay more for sustainable services, as well as that several companies that have their own sustainability programs do not usually inform the event organizers about them.

In this context, some interviewees mentioned that they sometimes contemplate the possibility of corporate sustainability concerns being more closely associated with the image companies wish to portray to their target audience rather than a genuine concern for the environment.

Concerning the level of corporate customers' interest in hybrid and online events, it is concluded that there has been a high increase in interest in online and hybrid events since COVID-19. However, after the pandemic, the interest in online events greatly decreased, and the importance of face-to-face events strengthened. Hybrid events continue to be relevant as a complement to face-to-face events to reach participants who cannot be present, reiterating the consistency of this trend in the future, as mentioned in the American Express Meetings & Events international study [7].

The interviewees also confirmed that internal meetings are the main type of event that takes place in hybrid and online formats, and incentives, team buildings, and celebrations are generally organized in a face-to-face format, as referred to by the author Nilsson [11].

About the evolution and change in the profile of corporate clients in the pre and post-COVID-19 pandemic timeframe, it was concluded that the main changes are associated with the increase of last-minute requests (causing the reduction of the event preparation period) and the demand for more flexible cancelation and reduction contractual clauses. There is also greater concern about the sanitary measures of the space and the region or country where the event is organized, as well as a greater demand for outdoor spaces, close to nature and with less tourism pressure. Additionally, was reinforced the current and growing importance of technologies, especially given the large adherence to the hybrid format, as a complement to face-to-face events.

It can therefore be concluded that the corporate customer profile in the Municipality of Cascais is defined by an increased interest in sustainability practices, although still at a very low level, and a current emphasis on hybrid events as a complement to face-to-face events. We also highlight the impact that the COVID-19 pandemic and the consequent restrictions had on the evolution and change of the profile, such as the increased importance of flexibility in the conditions for contracting services, concern regarding sanitary measures, and the demand for outdoor spaces.

As for the practical implications, it is important to highlight that a greater perception and awareness of the importance and application of diversity and inclusion measures by the organizers would represent an important improvement in the sector. It would also be relevant to make known the various existing sustainability programs and the respective progress in this area.

6. Study limitations and further research

Throughout this research, some limitations were encountered at the literature review stage as well as at the interview stage.

Regarding the literature review, the main limitation was the lack of studies and scientific articles that addressed the profile of the corporate events customer regarding sustainability trends, hybrid and online events, and their framework in the COVID-19 pandemic phase. Most of the information that could be obtained was within a general context of the events industry or the corporate universe but without a direct link to the COVID-19 period.

Regarding the interviews, the main limitation was the length of time it took to schedule the interviews, considering that the interviewees had limited availability, since the interviews took place during corporate events in high season months. Therefore, the interview period was about two months, for a total of 11 interviews.

To complete this research, we suggest that future research studies study the profile of the corporate client in larger geographical areas, such as Lisbon, or even at a national level. It would also be pertinent to make a more in-depth study of sustainability trends and hybrid and online events in the universe of corporate events, to obtain more detailed data and thus provide a greater academic contribution.

Conflict of interest

The authors declare no conflict of interest.

Author details

Cláudia Mendeiros[1] and Rita Peres[2*]

1 Estoril Higher Institute for Tourism and Hotel Studies (ESHTE), Lisbon, Portugal

2 Estoril Higher Institute for Tourism and Hotel Studies (ESHTE) and Centre for Research, Development and Innovation in Tourism (CiTUR), Lisbon, Portugal

*Address all correspondence to: rita.peres@eshte.pt

References

[1] Reic I. The development of the corporate events sector. In: Ferdinand N, Kitchin PJ, editors. Events Management: An International Approach. Los Angeles: Sage Publications; 2012. pp. 267-286

[2] Rogers T. Conferences and Conventions: A Global Industry. 2ª ed. Oxford: Butterworth-Heinemann; 2008

[3] Cavagnaro E, Postma A, Neese T. Sustainability and the events industry. In: Ferdinand N, Kitchin PJ, editors. Events Management: An International Approach. Los Angeles: Sage Publications; 2012. pp. 199-213

[4] Mair J, Smith A. Events and sustainability: Why making events more sustainable is not enough. Journal of Sustainable Tourism. 2021;**29**(11-12):1739-1755. DOI: 10.1080/09669582.2021.1942480

[5] Ton HNN, Le NK. Best practices for virtual events during the Covid-19 pandemic - focusing on attendee engagement. Advances in Social Sciences Research Journal. 2021;**8**(7):103-118

[6] Madray JS. The impact of COVID-19 on event management industry. International Journal of Engineering Applied Sciences and Technology. 2020;**5**(3):533-535

[7] American Express Meetings & Events. 2022 Global Meetings and Events Forecast. 2021. Available from: https://www.amexglobalbusinesstravel.com/meetings-events/meetings-forecast/ [Accessed: April 7, 2022]

[8] Davidson R. Business Events. 2ª ed. New York: Routledge; 2019

[9] Holmes K, Mair J. Event impacts and environmental sustainability. In: Page SJ, Connell J, editors. The Routledge Handbook of Events. 2ª ed. New York: Routledge; 2020. pp. 457-471

[10] Sox CB, Kline SF, Crews TB, Strick SK, Campbell JM. Virtual and hybrid meetings: A mixed research synthesis of 2002-2012 research. Journal of Hospitality & Tourism Research. 2015:1-40. DOI: 10.1177/1096348015584437

[11] Nilsson L. Hybrid Events: Breaking the Borders (Thesis). Traduzir: Lab University of Applied Sciences Ltd.; 2020

[12] Cassell C, Symon G. Qualitative Methods in Organizational Research: A Practical Guide. London: Sage Publications; 1994

[13] Dalfovo MS, Lana RA, Silveira A. Métodos Quantitativos e Qualitativos: Um Resgate Teórico. Revista Interdisciplinar Científica Aplicada. 2008;**2**(4):10

[14] Duarte R. Entrevistas em pesquisas qualitativas. Educar. Curitiba. 2004;**24**:213-225. DOI: 10.1590/0104-4060.357

[15] Vergara SC. Projetos e relatórios de pesquisa em administração. 12th ed. São Paulo: Atlas; 2010

[16] Roesch S, Becker G, Mello M. Projetos de estágio e pesquisa em administração: guia para estágios, trabalhos de conclusão, dissertações e estudos de caso. 2nd ed. São Paulo: Atlas; 1999

[17] Casimiro E, Gomes A, Almeida S. Plano estratégico de Cascais face às alterações climáticas – Sector Turismo. Câmara de Cascais. 2015. Available from: https://www.cascais.pt/sites/default/files/anexos/gerais/new/1_02_brochura_0.pdf. [Accessed: 2022-09-09]

[18] Barros BC. Cascais tem todas as condições para ser um destino de referência no turismo de negócios. Event Point, 08/06/2022. 2022. Available from: https://www.eventpointinternational.com/home/cascais-tem-todas-as-condicoes-para-ser-um-destino-de-referencia-no-turismo-de [Accessed: September 9, 2022]

[19] Visit Cascais. Resultados Inquérito, Turista de Cascais, Época Alta, Abril 2019 a Setembro 2019. 2019. Available from: https://associados.visitcascais.com/documentos/Observatorio/Estudos/Perfil-Turista/2019/Relatorio-Perfil-Turista-2019-Epoca-Alta.pdf [Accessed: September 11, 2022]

Chapter 14

The Impact of Pandemic Crisis on Hotel and Time-Sharing Accommodations in Greece

Konstantinos Varvaropoulos, Spyros Avdimiotis and Charalampos Vazouras

Abstract

This study aims at providing a comprehensive overview of the time-sharing accommodations and hotels, based on the view obtained from the pandemic health restrictions. Numerous areas were evaluated, including cancellations, rebooking, resumption of operations, hygiene measures, customer attraction practices, strategy adjustment and perception of competitors' activities, advantages and disadvantages. Focus is given on identifying the differences and similarities in the above areas, when comparing both forms of accommodation. Due to the nature of the subject and the lack of previous relative research, the qualitative research methodology is adopted. This has been achieved via the conduction of interviews, in a sample consisting of ten hotels and ten sharing-type accommodation businesses in various Greek regions. As for the results, cancellations were significant across the entire sample. However, hotels had a lower cancellation rate on average, as well as a higher level of rebooking, with almost all members in the sample reopening. Important disparities were observed in the approach to sanitary measures, as hotels achieved full consensus concerning their adoption and their necessity. Additional differences have been observed on marketing strategies, with the time-sharing accommodations being based more on price-competition. Finally, the contrasting strengths and weaknesses of each accommodation type are being examined.

Keywords: tourist accommodation, Greece, hospitality sector, sharing economy, pandemic crisis

1. Introduction

Recent pandemic crisis significantly affected everyday life, and a great number of businesses were not allowed to operate or operated under a special regime (e.g. servicing only take away) [1]. The purpose of this study is to record the reactions of host units to the effects of the pandemic of the SARS-CoV-2 virus, taking into account the measures preventing the spread of the virus [2]. This is because in addition to outlining how to deal with the effects of the crisis, it is quite possible that the practices may differ depending on the type of accommodation (traditional form such as hotels and sharing type like cottages).

IntechOpen

For Greek economy, tourism plays an important role as the participation of tourist sector in Gross Domestic Product of 2018 amounts exactly the 30.9%. Additionally, € 15.7 billion inflows into Greece from inbound tourism covering a significant part of the trade deficit [3].

Very little is known regarding the effect of pandemic on domestic hospitality sector. Research studies on the above effect are quite rare, as the lion's share belongs to e-learning [4, 5] and e-commerce [6, 7]. Consequently, the examination of the effect on hospitality industry, especially the comparison with sharing-type accommodations, is an unexplored scientific era. Through this research, cancelation and rebooking rates, new reservations, hygiene measures, pricing promotion policies, actions of competitors, and strategy realignment, both of hotel units and sharing-type accommodations, were evaluated.

Both accommodation types, hotels and sharing type, had distinct advantages and disadvantages against pandemic crisis, following different policies. Sharing economy accommodations were more flexible, while their pricing policy was not accurately predicted by hotels. Furthermore, both types of accommodation do not recognize each other as a direct competitor. Consequently, the current study comes to frame researches on hospitality industries of other countries [8–10], as well as to deal with sharing economy accommodations for the first time.

This paper is organized into four chapters. Unit 2 will provide the theoretical foundation concerning sharing economy and pandemic crisis of SARS-CoV2. Next section pictures the adopted research methodology, taking into consideration the research approaches and presenting the sample and selected research tool. In unit 4, a discussion of the survey results is described. Last but not least, the final chapter is a brief presentation of research findings and comments, limitations, and proposals of future research.

2. Literature review

2.1 Time-sharing economy

Since 2008, when the sharing economy emerged in the United States of America [11] as a concept and not as a practice, research efforts have been made on its impact on the economy. Given that the above practices extend to a great number of economic activities, such as hospitality and transportation, taking into account the multifaceted interactions it creates between units of the economy, economic or not, it is necessary to have a diverse and individualized approach.

Today, almost a decade later, it is viewed as an ever-growing economic, technological, and social phenomenon [12], with the current valuation of the peer-to-peer economy exceeding $75 billion [13]. The rapid expansion of sharing economy is due to the combination of two factors, technology and the financial crisis (Schor & [11].

The "sharing economy," as practice it is not something modern. It could be characterized as an instinctive trading practice [14]. The original concept used was "collaborative consumption," which was introduced by Rachel Botsman, but was soon replaced, as the term "sharing economy" prevailed in 2010 [11].

It is an undeniable fact that tourism has been greatly influenced by the practices of the sharing economy, as several aspects of it are directly related to areas where the sharing economy has developed rapidly. It is therefore easily understood that the tourism economy is the most appropriate field of research, as it combines different services in the tourist destination such as accommodation, transport, and catering [14].

Nevertheless, it should be noted that the spread of sharing in tourism is not the same across all services. To be more specific, a number of services are particularly widespread, while some others are relatively unknown. In particular, in a study by Stors & Kagermeier [15] in German cities, it emerged that while 1/3 of the tourist sample has made some kind of accommodation reservation, and only one in nine has received a free tour from a local.

In the sharing economy, there is a triangular relationship between the parties involved, i.e. the service provider, the recipient, and the relevant platform. However, in the case of tourism, two more parties are involved [14]. The first has to do with the government and market regulatory authorities such as the Hellenic Tourism Organization and the Independent Public Revenue Authority. The second category of stakeholders is none other than the traditional competitors, with the most typical example being hotels and rooms for rent.

In the context of dealing with the phenomenon and regulation of the sharing economy within the tourism economy, three different approaches have been proposed, prohibition, laissez-faire, and finally restriction [14]. The second approach is also the most flexible, imposing some additional fees on the accommodation. The approach of quantitative and/or qualitative restrictions is the most popular and has been proposed for implementation (with different limits over time) in Greece. More specifically, in an attempt to establish the first regulatory framework for the provision of accommodation within the sharing economy, the following restrictions were proposed but never implemented:

1. The property owner must be a person and not a corporation.

2. The permitted number of properties for short-term rental per owner is 2.

3. The maximum number of rental days is 90 per year, with the number being limited to 60 if it concerns islands with a population of less than 10,000 inhabitants.

4. The maximum income from short-term rentals must not exceed €12,000 per year.

5. The only provision is bed linen.

In the case of exceeding even one of the above limits, the exploitation ceased to be considered as operating within the framework of the "sharing economy," as a result of which it is treated as a business activity, with the main impact being the obligation to create a business, with all that entails (e.g. bookkeeping, VAT charge, etc.). In fact, it was planned to add additional criteria per region. However, this proposal has not been implemented to date.

2.2 SARS-CoV-2 crisis

The beginning of the recent pandemic crisis is located in 2019, in the city of Wuhan, Hubei Province in China, with the appearance of hitherto unknown cases of pneumonia [16]. The virus that was identified by the scientists was named SARS-CoV-2 and belongs to the wider family of coronaviruses, such as SARS and MERS. The disease was named COVID-19 and despite worldwide efforts to reduce it and quickly spread worldwide [16].

In the recent past, tourism industry has been significantly affected by a wide range of crises such as terrorism (9/11), weather events (e.g. Hurricane Katrina), and epidemics (e.g. SARS) [17]. Regarding the 2004 tsunami, Henderson [18] estimates that 67.2% of arrivals fell in the first half of the following year, a decline that led to the collapse of 500 businesses and the loss of 3000 jobs. Focusing on the case of SARS, the ancestor of the recent virus, it was estimated that 3000,000 workers lost their jobs in Asia, while China, Hong Kong, Vietnam, and Singapore lost more than $ 20 billion of GDP [19]. At this point, it is crucial to mention that COVID pandemic was the reason for the largest decrease in international tourist arrivals since World War II. The parallel of the recent pandemic with SARS presents significant weaknesses, as its impact on the tourism industry could not be considered global.

The hotel industry was strongly affected and was immediately called upon to face the unexpected challenge [20]. For instance, in the beginning of the pandemic, hotel occupancy in China Mainland dropped from 70% to just 17% within 2 weeks [21]. According to the Hellenic Chamber of Hotels [22], the reduction in the number of bookings compared to 2019 ranges from 72–92% to 58–83% in 12 months and seasonal hotels, resulting losses more than half billion euros. Moreover, within the above framework, jobs are expected to decrease by 38,234.

Nevertheless, the emergence of a crisis, in addition to the adverse consequences, could also have beneficial effects on tourism industry. More specifically, it strengthens domestic tourism, facilitates modernization, disseminates digital transactions [23], and finally, could provide an opportunity for a destination to be restructured, transformed, and presented with a new alternative tourism proposal (e.g. saturated destinations have exceeded their carrying capacity).

3. Research methodology

3.1 Research approach

Two different research approaches have prevailed in the scientific community, which represent two different approaches. From a chronological point of view, there is the quantitative approach which is also known as positivism, while the second is the qualitative approach and is also known as interpretivism [24].

According to Kvale [25], the most important difference between the above methods lies in the fact that while positivism adopting a quantitative perspective calculates the frequency of observations to lead to the formulation of laws, interpretivism as a qualitative methodology analyzes each phenomenon separately, giving emphasis on its unique features in order to formulate broader conclusions.

At this point, it is useful to mention that at the end of the 20th century, the third alternative research methodology was developed, criticism. This methodology, attempting to address the weaknesses of each existing method, tries to combine their elements. More specifically, according to criticism, the data can be both collected and created, while they are then processed through a repeated statistical model, so that through tests the research hypothesis can be confirmed or rejected [26].

As for the selected research methodology, interpretivism and the qualitative approach are the ideal solution. Initially, criticism, although accompanied by strong optimism, is a relatively modern method with limited empirical application. This fact makes his choice both problematic and dangerous. From then on, the modern nature of the research object and the absence of previous research lead to the rejection of positivism and the quantitative approach. Consequently, through interpretivism, the initial approach to the object of research is attempted in order to understand the characteristics of the phenomenon and formulate inductive generalizations.

3.2 The sample

In the case of the present research, the population consists of all hotels and sharing-type accommodations within the Greek territory. Taking into consideration that gathering data from the entire population is impossible, the characteristics we are interested in will be collected from a part of the population, which is called sample. In fact, these characteristics are in the research variables [27].

The research sample consists of 10 hotel units of various stars and 10 sharing economy accommodations. For each accommodation, either the owner or the manager will be asked.

Given that specific individuals from the hotel industry and sharing accommodations were approached for the research, the sampling is characterized as purposive. The selection of a sample from a network of acquaintances rests on two axes. The first has to do with creating a climate of trust in order to collect as detailed and reliable data as possible. The second has to do with unfavorable timing and the possibility of nonresponse. Finally, the sample is presented in detail in the appendix.

3.3 Research tool

As already noted above, the qualitative research approach is adopted. The research tool chosen is that of the structured interview. The advantages provided by the structured interview for the implementation of this research can be presented as follows [28]:

1. It allows respondents to speak in their own words about any issue that is important to them.

2. It can be insightful as it provides causal inferences.

3. It propels the rise of unexpected data.

4. It allows observation of elements such as tone of voice.

5. It allows the gathering of inside and confidential information due to personal contact and the eventual development of a climate of trust.

A comparative advantage of the interview over other research tools is that the interview, beyond the control of the research case, may indicate new hypotheses, while supporting the identification of variables and connections, which may not have been taken into account [29].

More specifically, through the structured interview, the questions are subsequently determined but provide the researcher with the opportunity to deviate partially during its process. This is based on the fact that it provides the necessary freedom, but within a predetermined framework [29]. In this way, on the one hand, the open approach to the research question is combined, while avoiding the risk of disorienting the interview.

The interview questions include the descriptive image of the sample, objective data regarding the interviewee's job position, subjective data such as opinions, and finally behavioral data regarding actions within specific circumstances. The ultimate goal of interviews is to allow the sample members to develop their positions and articulate their concerns and beliefs.

4. Results

4.1 Cancelation of bookings

Both at hotel units and sharing-type accommodation, cancelations were almost total. To be more specific, cancelations in 4 of 10 hotels in the sample examined were total, while 1 hotel unit did not have a reservation until March; therefore, it was not affected. In the rest of the hotels, cancelations ranged from 50 to 90%, with the lowest percentage recorded in two units on the island of Tinos. In all cases, the hotels' remaining reservations concerned individual travelers and mostly Greeks, while tour operators canceled all their reservations.

As for sharing-type accommodation, the situation was no better. Almost the entire sample experienced cancelations of 80–100% of summer bookings. In fact, the cancelations took place massively with the outbreak of the pandemic, as they took place within just 10–15 days of March. An exception was an accommodation in the Regional Unit of Lasithi, which recorded cancelations of around 10%, while in an accommodation in the Regional Unit of Halkidiki there was no reservation until the outbreak of the pandemic.

The devastating effect of the pandemic on the hospitality sector is a fact, as it led to the cancelation of almost all bookings. More hotels lost 100% of their bookings in relation to sharing-type accommodation, while bulk bookings by tour operators were canceled in their entirety. Characteristic of the remaining bookings in both types of accommodation is that they concerned individual tourists in the overwhelming majority from within the country.

4.2 Rebooking

The tendency to rebook seems to vary significantly between hotel units, with rebooking ranging from 0 to 40%. In contrast, the range of rebooking in sharing-type accommodation is lower.

In particular, two hotels in the sample, located in the city of Trikala, despite the fact that they had returned the advance payments from the beginning and considered the reservations as lost, saw about one in three cancelations come back. This fact is largely due to business customers and not tourists. On the other hand, in the two hotel units on the island of Tinos and one in Elati Trikala, no cancelation was "revived." Regarding the reservations of two hotel units located on the island of Skiathos, there were some that moved booking dates to later. In the units of the Regional Unit of Thessaloniki, rebookings were minimal, exclusively of individual travelers and did not exceed 10%.

In sharing-type accommodations, there has been some rebooking interest, but repossessions have either not been completed or have been minimal. In the entire survey sample, rebooking did not exceed 10% of cancelations, while some of them were either moved to autumn or summer of 2021. It is again worth noting that most of the rebookings concerned the accommodation located in the Lassithi Regional Unit, which recorded the fewest cancelations.

With numbered exceptions, the majority of the sample, regardless of type, did not detect a significant trend for rebookings. More specifically, on a percentage basis, rebooking in 18 of the 20 accommodations did not exceed 10%.

4.3 Reservations after the end of restrictions

The lifting of restrictions increased bookings as expected, but not satisfactory for both hotel units and sharing-type accommodations.

Only in one of the hotels in the sample, bookings did not show significant differences from previous years. Nevertheless, it should be noted that this is a mountain hotel, outside of the high tourist season, and therefore the bookings were very few anyway. From then on, almost the entire sample agrees that proportionally there was rudimentary demand, but this is due on the one hand to the compression of the season and on the other hand that from the zero bookings that were left, achieving any occupancy is a significant increase. The bookings of most of the hotels in the sample were not enough to generate profits, as the units were operating at break-even at best. It is important to point out that in two hotel units on the island of Tinos, although the bookings were less than in previous years, they were considered satisfactory, recording to a certain extent an unexpected replacement of some of the losses of older, possibly more cautious customers, by younger ones.

Regarding sharing-type accommodation, there was interest, but this rarely transformed into bookings. In fact, although in percentage terms bookings increased significantly after the lifting of the measures, in absolute terms and compared to the corresponding months of 2019, the situation was clearly more unfavorable. The smallest deviation compared to previous years was recorded again in the accommodation of the Lassithi Regional Unit. It should be noted that the course of bookings was associated with borders' opening and in the case of Halkidiki with the land borders. At the same time, there was a tendency for last-minute bookings, which for most accommodations was unprecedented.

Taking the above into consideration, it is obvious that after the lifting of the measures, there was a clear upward trend in bookings, which, however, did not approach the figures of earlier years and is not considered satisfactory by the vast majority of accommodations, regardless of type.

4.4 Opening after the lifting of measures

The majority of the accommodations in the sample were made available to the public with the lifting of the restrictions.

Nine of the 10 hotels opened normally and depending on whether they are seasonal or 12-month accommodation. Just one hotel in the Regional Unit of Thessaloniki decided to remain closed for the entire summer. However, it should be noted that in this particular case, the management of the unit decided to proceed with a partial renovation, while any reservations would be transferred to a unit of the same business group without causing any problem.

Regarding the opening date, all sharing-type accommodations were available to the public from the first day the suspension was lifted. In fact, the accommodations were available even if they did not have a reservation to serve immediately.

It is obvious that essentially the whole sample, regardless of whether they are hotels or sharing-type accommodations, were available from the first day after the restrictions' period. The one exception that was recorded and concerns a hotel is not solely due to limited bookings but to taking advantage of the situation and treating it as an opportunity for renovation.

4.5 Additional hygiene measures

Additional hygiene measures are an important area of differentiation between hotels and sharing-type accommodations.

First of all, as a matter of fact those hotels were obliged to implement a specific health protocol, they were asked to take additional measures. However, the majority of hotels stated that several of the protocol's measures were already in place, while almost all of them voluntarily took additional measures. Apt examples were the exclusive use of cleaners approved by the National Organization of Medicines, the more regular change of substrates and the higher than the specified washing temperature of bed linen, the use of UV technology sterilizers and fogging disinfection machines and the personal protection equipment of the staff. At this point, it is vital to note that there have been reports of other hotels partially deviating from the protocol due to customer pressure (how breakfast is served and frequency of cleaning).

On the side of sharing-type accommodation, significant differences are observed. The majority of accommodations used stronger, professional cleaners and did more thorough cleaning, but without anything specialized. Two properties took several additional measures such as using a steam cleaner and sterilizer, providing gloves, masks, and disinfectants to customers. In fact, one of them tried to optionally implement the hotel protocol. On the other hand, three accommodations did not implement any additional measures. Finally, it should be noted that a significant number of accommodations either provided the possibility or implemented check-in and check-out procedures without personal contact. In addition, some did not allow consecutive bookings, leaving the accommodation empty for a few days.

Considering the above, it is more than obvious that hotel units proceeded in taking more measures compared to sharing-type accommodations. In fact, most hotels were not satisfied with observing the mandatory governmental protocol but also took additional measures to protect the health of customers, employees, and the reputation of the company. On the contrary, and given the absence of a specific health protocol, no special measures were applied to sharing-type accommodations, apart from two exceptions that worked in the same contexts as hotels, while three of them did not receive any additional measures compared to the summer of 2019.

4.6 Perceptions on the existence of mandatory measures

Regarding the existence of mandatory health measures such as the hygiene protocol for hotels, a more general acceptance of them is recorded. However, acceptance seems to stem from different factors and have different intensity.

All hotels have a positive attitude toward the existence of a sanitary operating protocol. In fact, a significant number of the sample stated that some of the measures were observed anyway, while the existence of additional measures such as measuring the temperature of customers upon entering the hotel unit was suggested. The vast majority of hotels seek to maintain part of the health protocol even after the end of the pandemic, as despite the cost and time, it creates value for the industry, protects employees and customers, and differentiates professionals from opportunists in the industry. On the contrary, there have been complaints from customers, especially in the area of serving breakfast. The question of the universality of the application of the measures was also raised, as their noncompliance by other hotel units increases the discomfort of customers and creates an issue of unfair competition.

Sharing-type accommodation is also moving positively toward the existence of measures to protect health. However, two accommodations seem to be unaware of the details of the existence of a health protocol, possibly because it does not directly concern them, while two others treat them neutrally and possibly more strictly than they should. A property considers them necessary due to the situation but considers that since there are no consecutive bookings, their feasibility is limited. On the other hand, almost half of sharing-type accommodations in the sample deal with the existence of a hygiene protocol with greater interest, taking its existence for granted. In fact, it is emphasized that the existence of a sanitary framework should have already been provided by accommodations themselves and that it should not have been imposed by the government. The accommodations that are more interested in the existence of sanitary measures cannot be geographically related, as one is located in the Regional Unit of Lasithi, one in Thessaloniki, and two in Halkidiki.

The above results could be justified analytically. The sample absolutely agrees on the adoption of health measures. Hotel units that operate as corporations show greater interest in the existence and observance of hygiene protocols, as they believe that in this way they are upgrading the quality of the industry's services and protecting their own businesses. On the other hand, sharing form accommodations, given that they are not required to adopt any protocol and most of them do not operate with purely business criteria, are presented more neutrally. However, it should be underlined that a significant number of them recognize the necessity of the measures, both from a health and reliability point of view.

4.7 Price reduction

Reducing the nightly rate as a tool to attract customers is another point of differentiation between hotels and sharing-type accommodations.

The majority of hotel units in the sample stated that they will not reduce their prices. The above decision is based on the increase in costs brought about by the observance of the hygiene protocol but also on the already low prices (either as a result of the crisis or due to the point of the season). Instead of reducing prices, hotels seem to focus on different types of policies to attract customers. As for the hotels that will proceed with price reductions, they are all located in island areas, two on the island of Tinos and two on the island of Skiathos. The first ones adopt a price reduction of up to 20%, during the first period of opening and exclusively for the cheapest rooms, leaving the suites and seaside rooms at the same prices as 2019. On the contrary, on the island of Skiathos, the price reduction approaches 50%, recognizing that this decision contradicts the rest of the competitors, seeking increased occupancy and low-profit margin.

Regarding sharing-type accommodation, the vast majority (7 out of 10) reduced their prices, which in most cases ranged between 10% and 20%. One accommodation that did not reduce prices, based on Lassithi Regional Unit, focused on last-minute discounts and special offers in order to limit the days when the accommodation will remain empty. Finally, only two accommodations in the Halkidiki Regional Unit did not proceed with any price reduction, considering their prices to be low anyway and demand to be negligible.

The above results show a clear indirect or direct decrease in prices in both types of accommodation. The price reductions in hotel units were made mainly in accommodations located on islands, while in some cases they were not universal for all rooms. On the other hand, almost all sharing-type properties saw a small or slightly larger decrease in prices. The above differentiation is estimated to be linked to the inelasticity of fixed costs of the hotels (e.g. rents and labor) as well as the cost of implementing the health protocol. The above factors do not exist in sharing economy accommodations, thus allowing for an easier reduction in nightly rates.

4.8 Other customer attraction actions

As it has already mentioned, price reduction is a point of differentiation between the two different types of accommodation. In fact, the inelasticity of prices in hotel units is usually accompanied by other methods to attract customers.

To be more specific, considering the impossibility of a significant price reduction, the majority of the hotels in the sample implemented new methods of attraction. Apt examples are special offers such as late check-out and the addition of a free night for bookings greater than a minimum limit (e.g. for three nights an additional one at the same price). A number of hotels focused on the flexibility of reservations and the provision of facilities to change them as well as the highlighting of health measures. Most of the four- and five-star accommodations focused on achieving direct bookings through direct domestic advertising, while one hotel attempted to enter new markets by participating in public tenders with minimal profit to achieve occupancy and by displaying on sharing-type accommodation platforms. Finally, one hotel unit limited the promotion abroad as on the one hand it considered it pointless and on the other hand there was fear for the safety of the remaining customers and employees.

Regarding sharing economy accommodations, the great majority, limit themselves to price reduction and do not adopt alternative customer attraction practices. More specifically, only three properties focused on more intensive use of social media (e.g. Facebook Marketplace) and one went for last-minute promotions. A significant number of properties considered demand to be limited anyway, resulting in no reason to advertise, while in several cases properties stated that they do not operate purely as a business and do not apply other marketing tools beyond price, image, and positive comments on the platforms.

Taking the above into consideration and in combination with the previous subsection, it is clear that hotel units focused more on alternative methods of attracting customers (e.g. special offers and participation in competitions), while sharing-type accommodations operating less entrepreneurially is limited to lower prices and accept the loss of this year's season more easily.

4.9 Perception of competitors' actions

The perception of the movements of the competition is a key factor in the formulation of corporate strategy. In the present case, accommodations appear to have a relatively accurate picture of the competition, with the exception of hotel units that expected significant reductions in shared accommodation prices, but this was not proven by the results.

Most hotel units consider that their direct competitors (other hotels) either did not reduce prices or reduced them only slightly. However, there are some exceptions, as there are hotel units with many stars or specific areas (e.g. Sporades), where the reduction in prices was more intense. However, the majority of hotel owners and managers agree that the reduction in prices should not have paid much and could be more a panic move than a strategic choice. At the same time, it is believed that several four- and five-star hotels have chosen not to operate.

Regarding sharing-type accommodations, there is a wider picture that has put the weight on reducing prices by lacking firm policies, following a more aggressive and possibly opportunistic policy. At this point, however, it must be emphasized that the majority of the sample of hotel units do not consider sharing-type accommodations as direct competitors, while they believe that hotels are the regulators of market prices. On the contrary, the owners and operators of sharing-type accommodations consider that they are competitors of "rooms for rent" due to the nature of their "product."

Sharing-type properties claim that hotels have made smaller or even no price reductions, while some have decided to remain close. They even claim that hotel units are based on cleanliness and hygiene as well as special offers. They also believe that several hotels turn to repeat bookings of past customers taking advantage of the existing relationship and some attractive offer.

As for sharing-type accommodations' view of competition among themselves, antagonism on price per night is assumed, as this is afforded due to cost. On the other hand, there is a tendency to limit the supply of this type of accommodation, either due to their change to long-term leases or due to owner-occupation.

Taking the above into consideration, it is accepted that hotel units did not proceed with significant price reductions, especially compared to sharing economy properties. But, despite the belief that the second one will proceed with large price reductions, the results differ. It is therefore quite possible that the opinion prevails both among hotels and among sharing-type accommodations about large reductions in the prices of sharing economy rents, but an opinion which is not confirmed in practice by the sample. Also, the sharing-type accommodations claim that there is a decrease in supply, which in this case is confirmed by their own practices (e.g. increased owner-occupancy and long-term renting).

4.10 Strategy realignment

Both hotels and sharing-type accommodations proceeded to readjust their strategy. Nevertheless, strategy change seems to differ.

For the majority of the hotels, the change in strategy is connected with the implementation of the hygiene protocol in their wider operation, while a significant number of them reduced costs in order to make their operation sustainable. It should also be noted that the hotels turned their attention to direct bookings and especially to domestic tourism as well as strengthening the visibility of their corporate identity, with an emphasis on health and cleanliness. There were also cases of more intense changes such as the interruption of operations, the service of "relative" hotel units that did not open after the end of restrictions, participation in public tenders, and presence on sharing economy platforms.

In terms of strategy, the great majority of sharing economy properties made notable changes. Firstly, one in three properties stated that their summer strategy was not static but was changing according to recent conditions. Focusing on the changes, three accommodations increased the period of summer expropriation, since they considered the given situation to be an opportunity for alternative exploitation of their property. On the other hand, two accommodations in the Regional Unit of Halkidiki chose the lease for a medium-term period (1 and 2 months), while a third in the Regional Unit of Thessaloniki turned to an annual lease, in order to return to the short-term rental market after the end of the pandemic crisis. Smaller changes in strategy concern pricing policy, more flexible booking policy, and visibility (strengthening direct targeted online visibility).

From the above, it can be concluded that pandemic crisis brought vital changes in accommodations' strategy. However, changes differ significantly between the two types of accommodations. The changes in the hotels are linked to the mandatory implementation of the hygiene protocol, the change of the target market, while there are also important changes such as the suspension of corporate operation. On the contrary, sharing-type accommodations focused on pricing policy and different use of the property, something that could not happen in hotels. A common component, however, is the emphasis on direct domestic bookings.

4.11 Competitive advantages: disadvantages

Regarding the competitive advantages and disadvantages of hotels and sharing-type accommodations, there is a consensus among them.

Hotel units consider as their comparative advantages the existence of a hygiene protocol, their guaranteed application due to the controls carried out, the sense of professional handling of cleanliness and hygiene (even if this means refusing a reservation), respect for the customer, and the reputation. A number of hotels also mention the existence of prices and offers that make the stay financially feasible, compared to previous years.

On the other hand, the weak points that hotels recognize in relation to sharing-type accommodation are the mandatory contact with foreign people (employees and other residents), the entry of a third party (cleaning lady) into the room, the inability to compete on price due to costs, most restrictions (e.g. wearing a mask in public areas) as well as the impossibility of complete isolation.

Following the mentioned trend, sharing economy accommodations consider as their competitive advantage the possibility of complete isolation of guests and the lack of contact with third parties. There is no interaction with employees and other residents, and there are no common areas with the risk of overcrowding, while residents can even dine in a space fully controlled by them or take advantage of specific amenities such as a private beach. Within the above context, part of the sample recognizes as an advantage the possibility of renting for a longer period (e.g. a month or the entire summer season), offering a "different product" to customers who want their "own" holiday home.

On the other hand, as disadvantages seems to be the reduced resources and the lack of specialized staff (e.g. cleaners), the absence of any hygiene protocol sets a common denominator for everyone, limited services (e.g. breakfast in a box), the uncertainty regarding the level of cleanliness and disinfection, and finally the lack of economies of scale which limits the possibilities of accommodation (e.g. it is unprofitable to purchase specialized disinfection machines for the operation of only one house).

4.12 Additional comments

There are two main concerns of hotel units regarding the recent pandemic. The first has to do with the attitude of the state in the event of a case of infections despite the observance of all the required measures. The protection of the hotel's reputation and its support in a possible suspension of its operation are important issues for a hotel unit which, despite the adoption of measures even stricter than the protocol, is faced with virus cases. A wider fear was even observed, for the health of the workers and the reputation of the company.

The second concern is related to the existence of behaviors contrary to the imposed protection measures. The discomfort of customers for specific measures such as the way breakfast is served or the additional services expected from a high-star hotel is amplified in case of experiences in corresponding hotels that circumvented the measures. Consequently, the existence of a hygiene protocol turns from an advantage of the sector into a tool of unfair competition, and there is an underlying risk of deregulation of the sector's market.

One point of view that may need more extensive study is the geographical factor of the destination. It has been underlined by an accommodation as a factor influencing bookings, pricing policy, or attraction methods. This observation may have some basis if we consider that some destinations rely on foreign tourism (e.g. northern Greece and visitors from Balkan countries), while access to other destinations could not be achieved exclusively by a private car but requires the use of "dangerous" for health public transport (e.g. islands accessible by planes and ships).

5. Conclusions: limitations: proposals

5.1 Conclusions

The results of the survey, as presented in the previous section, lead to some interesting conclusions regarding the recent pandemic crisis and how hotels and sharing-type accommodations were dealing with it.

First of all, summer bookings were almost entirely canceled during March. In the great majority of sharing-type accommodation, cancelations ranged between 80% and 100%, with only one exception (10% in Lassithi Regional Unit). On the contrary, in terms of hotels, more cases of cancelations of all reservations were recorded. Furthermore, significant fluctuations were observed in the rest sample, with some hotel units having lost only half of their summer reservations (two hotels on the island of Tinos). In general, hotels that mostly cooperated with tour operators were more exposed to the risk of cancelations, while most of the remaining bookings made from individual domestic tourists. This is probably also the reason why sharing economy properties, despite the fact that they had the majority of higher cancelation rates, did not experience total cancelations.

Regarding rebookings, hotel units recorded on average higher rebooking rates, which in certain cases approached 40%. Conversely, rebookings fluctuated at lower levels for sharing economy accommodation. The above can be explained as the highest rebookings were presented in city hotel units and concerned business customers and not tourists.

Bookings after the end of limitation measures, especially in some areas such as Halkidiki Regional Unit, fluctuated at low levels inextricably linked to the opening of the land borders. The only cases where the course of bookings was deemed satisfactory were the hotels on the island of Tinos and sharing-type accommodation in the Regional Unit of Lasithi, accommodation which also recorded the lowest percentage of cancelations. This fact possibly reveals a differentiation of the effect of the pandemic according to the geographical location of the destination.

Despite the existence of significant cancelations, only one hotel from the entire sample did not open after the limitations were lifted, with the great majority of properties taking advantage of the remaining season. But it should be noted that the only case of accommodation that was not put into operation concerns a case of renovation, having already found a way to serve its customers.

In the field of health measures, the hotel units showed greater homogeneity in their responses, most obviously due to the existence of the hygiene protocol. In fact, in most of the hotels in the sample, some meters of hygiene protocol were already being implemented. The professional treatment of accommodation seems to have played a vital role, as these are businesses that wish to protect their reputation. On the other hand, in sharing-type accommodation, significant differences in measures were observed. The absence of any meter as well as the different treatment of the accommodations by the owners-managers led to significant discrepancies. In particular, there were sharing-type accommodations that took measures parallel to those of the hotel hygiene protocol, but also accommodations that did not take any additional measures compared to previous years.

Focusing on hygiene protocol, hotel units were strongly positive about its implementation, with some of them wishing not only to take additional measures but also to indefinitely extend its implementation. Nevertheless, there was also concern, as incidents of selective application of the protocol to competitors were recorded. These incidents lead to unfair competition and nullification of any advantage the protocol offers to the industry. On the contrary, sharing economy accommodation also showed a general positive attitude toward the adoption of a protocol, however, with significant differences. Some accommodations consider the existence of a protocol necessary for them as well in order to upgrade the "industry," while some others present themselves more neutrally.

An important area of differentiation between hotels and sharing-type accommodations is pricing policy. The majority of hotels did not proceed with any price reduction, as the increase in operating costs due to the protocol combined with the decrease in demand made it prohibitive. Any small declines were mainly related to specific room types and the period immediately after the restrictive measures were lifted. The greatest price reductions were recorded in islands (Tinos up to 20% and Skiathos up to 50%) and are likely to be linked to accessibility and customer mix (individuals and tourist packages). On the other hand, almost all sharing-type accommodations decreased their prices between 10% and 20%. Obviously, the greater elasticity of costs in sharing accommodation has enabled owner-managers to proceed with price reductions.

The reluctance of hotels to reduce prices seems to propel them to adopt different ways of customer attraction such as special offers. More specifically, hotel units chose to implement special offers such as late check-out, greater flexibility in bookings, and the offer of additional free nights, depending on the days of initial booking. At the same time and realizing the trend as reflected in the remaining bookings, the hotels focused their advertising within the country, targeting individual bookings. On the other hand, sharing economy accommodations seem to have limited themselves to price reduction, while a small number of them increased social media usage. It could be said that sharing economy properties accepted this year's tourist season as lost, reduced prices to get what they could, unlike hotels which characterized by a more professional approach.

Regarding antagonists' actions, hotel units assumed, and it seems correctly that their direct competitors (e.g. other hotels) will not proceed with price reductions or will only slightly reduce them. In fact, even for the cases where there were large reductions, they believe that they are panicked and took ineffective decisions. On the contrary, although they do not face sharing-type properties as direct competitors, they expected, wrongly as it appears from the results, significant reductions. On the other hand, sharing-type accommodations assumed that competition with other accommodations of the same type will be based on price. As for hotels, they recognized that they have limited possibilities to reduce prices, as well as a considerable number of them, especially seasonal, will not operate. Nevertheless, last one was not verified by research.

Changes in strategy were also clearly observed, which were even more intense in the cases of sharing-type accommodation. Hotel units focused on hygiene, cleanliness, and corporate identity, while there were also some more radical changes such as the presence on sharing platforms and participation for the first time in public tenders. In contrast, several sharing form properties have completely changed their strategy. The above changes achieved by entering into monthly and longer rents or by increasing owner-occupancy not only by the owners themselves but also by relatives. It is obvious that the flexibility of operation of sharing economy accommodations compared to hotels, and the more "opportunistic" operation of the majority of them, led to the vital changes mentioned above.

Finally, regarding the comparative advantages of hotels, these are the existence and guaranteed implementation of the hygiene protocol, the professional approach to cleanliness, the available resources, the reputation, the customers' sense of security, the respect he receives, and the existence offers that were not provided in the past. On the other hand, sharing-type accommodations in the midst of a pandemic had to demonstrate competitive prices, the possibility of complete isolation, flexibility in renting (e.g. renting for the whole summer and using it as a family holiday home) as well as greater comfort (use of common areas without restrictions such as the garden or the private beach).

5.2 Limitations

The limitations of the research are directly related to the chosen methodology and the research tool of interviews.

Firstly, random selection of the sample through existing personal contacts, in order to achieve the maximum response rate and its relatively limited number, are quite likely not to accurately represent the population. For instance, most shared accommodation is located in the Halkidiki Regional Unit. On the other hand, hotels are located in the Regional Units of Thessaloniki, Trikala, Magnesia, Sporades, and Cyclades. From the above, it is obvious that maximum possible dispersion of the sample is not achieved.

The second limitation has to do with remote interviewing. Although the teleconference method was preferred, where this was not possible, the interview was conducted via telephone call. This limited the interviewer's ability to detect reactions such as body posture, which would have been an apt evident via live contact.

Finally, the low assessment of their role by a significant part of the owner-managers of sharing-type accommodations led to more limited time interviews compared to hotel owner-managers. The above is explained, given that a view of less coordinated and organized operation was identified. As a consequence, answers are concise and there is a lesser degree of their documentation and interaction.

5.3 Proposals

Taking into account the above, it is obvious that the current research constitutes an initial basis for further investigation of the subject. More specifically, the research is proposed to be extended to accommodations of regional units beyond the selected ones, enlarging the sample. In this way, a more complete picture of the entire Greek territory will be achieved.

Furthermore, as the amount of the price reduction of sharing economy accommodations is lower in the sample than the recorded perception of hotels, it is proposed to carry out a quantitative research on the size of the price reduction due to the pandemic in sharing-type accommodation, in order to examine the hypothesis and to confirm or reject the formed perception. It would be particularly useful for hotel sector to understand more effectively the pricing policy of sharing accommodations, even if they do not consider them as direct competitors.

In addition to the above, based on the responses collected from hotels and the significant variation in price reduction, it is suggested to investigate whether there is a correlation between rate of price reduction and factors such as geographic region, seasonality, and customer mix (e.g. individual or massive). This as factors like means of transport is likely to affect the demand for specific destinations. Moreover, it could be helpful for hotel sector to evaluate the effect on prices of reliance on intermediaries (e.g. tour operators).

Finally, as health crisis remained in the summer of 2021, it would be interesting to repeat the survey, with the accommodations having already gained valuable experience during the summer of 2020 and possibly having decided to readjust their practices.

A. Appendix

Hotels participated in the research and make up the corresponding sample are the following:

1. La Piscine Exclusive Art Hotel ***** in Skiathos island.

2. Skiathos Avaton Hotel**** in Skiathos island.

3. Imperial Palace**** in Thessaloniki.

4. Imperial Plus**** in Thessaloniki.

5. Panhellenion*** in Trikala.

6. Trikala River House*** in Trikala.

7. Aktaion Hotel*** in Afissos, Magnesia.

8. Posidonio** in Tinos island.

9. Venus Minimal Hotel** in Tinos island.

10. Mikri Arktos** in Elati, Trikala

Accordingly, the 10 shared short-term rental accommodations are as follows:

1. *Room11* in Thessaloniki.

2. Scandinavian #102 in Thessaloniki.

3. Ground floor country house in Geoponika, Halkidiki.
4. First floor country house in Geoponika, Halkidiki.
5. Beach House in Siviri, Halkidiki.
6. Cozy Apartment in Poseidi, Halkidiki.
7. Apartment with magnificent sunset view in Hanioti in Hanioti, Halkidiki.
8. Achinos Luxury Accommodation in Pefkohori, Halkidiki.
9. Traditional Stone House in Kotronas, Laconia.
10. Beachfront Villa Kirvas in Koutsounari, Lasithi, Crete island.

Author details

Konstantinos Varvaropoulos*, Spyros Avdimiotis and Charalampos Vazouras
International Hellenic University, Thessaloniki, Greece

*Address all correspondence to: k.varvaropoulos@outlook.com

References

[1] Bartik A, Bertrand M, Cullen Z, Glaeser E, Luca M, Stanton C. How are small businesses adjusting to COVID-19? Early evidence from a survey. HKS Working Paper No. RWP20-012, University of Chicago, Becker Friedman Institute for Economics Working Paper No. 2020-42. DOI: 10.3386/w26989

[2] Haug N, Geyrhofer L, Londei A, Dervic E, Desvars-Larive A, Loreto V, et al. Ranking the effectiveness of worldwide COVID-19 government interventions. Nature Human Behaviour. 2020;**4**(12):1303-1312. DOI: 10.1038/s41562-020-01009-0

[3] Ikkos A, Koutsos S. Η συμβολή του Τουρισμού στην ελληνική οικονομία το 2018 [The contribution of Tourism in the Greek economy in 2018]. Athens: The Greek Tourism Confederation (S.E.T.E.); 2020

[4] Maatuk A, Elberkawi E, Aljawarneh S, Rashaideh H, Alharbi H. The COVID-19 pandemic and E-learning: Challenges and opportunities from the perspective of students and instructors. Journal of Computing in Higher Education. 2022;**34**:21-38. DOI: 10.1007/s12528-021-09274-2

[5] Ebner M, Schön S, Braun C, Koschutnig-Ebner M, Grigoriadis Y, Haas M, et al. COVID-19 epidemic as E-learning boost? Chronological development and effects at an Austrian university against the background of the concept of "E-learning readiness". Future Internet. 2020;**12**(6):94-113. DOI: 10.3390/fi12060094

[6] Jílková P, Králová P. Digital consumer behaviour and eCommerce trends during the COVID-19 crisis. International Advances in Economic Research. 2021;**27**:83-85. DOI: 10.1007/s11294-021-09817-4

[7] Susmitha K. Impact of COVID 19 on E-commerce. Journal of Interdisciplinary Cycle Research. 2021;**12**(9):1161-1165. DOI: 10.13140/RG.2.2.20236.85127

[8] Burhan M, Salam M, Hamdan O, Tariq H. Crisis management in the hospitality sector SMEs in Pakistan during COVID-19. International Journal of Hospitality Management. 2021;**98**:1-12. DOI: 10.1016/j.ijhm.2021.103037

[9] Rodriguez-Antón J, Alonso-Almeida M. COVID-19 impacts and recovery strategies: The case of the hospitality industry in Spain. Sustainability. 2020;**12**(20):1-17. DOI: 10.3390/su12208599

[10] Karim W, Haque A, Anis Z, Ulfy M. The movement control order (MCO) for COVID-19 crisis and its impact on tourism and hospitality sector in Malaysia. International Tourism and Hospitality Journal. 2020;**3**(2):1-7. DOI: 10.37227/ithj-2020-02-09

[11] Schor J, Cansoy M. The sharing economy. In: Wherry F, Woodward I, editors. The Oxford Handbook of Consumption. Oxford: Oxford University Press; 2019. pp. 51-74. DOI: 10.1093/oxfordhb/9780190695583.001.0001

[12] Hamari J, Sjoklint M, Ukkonen A. The sharing economy: Why people participate in collaborative consumption. Journal of the Association for Information Science and Technology. 2015;**67**(9):2047-2059. DOI: 10.1002/asi.23552

[13] Allen D. The sharing economy. Institute of Public Affairs Review: A Quarterly Review of Politics and Public Affairs. 2015;**67**(3):24-27

[14] Avdimiotis A, Poulaki I. Airbnb impact and regulation issues through destination life cycle concept.

International Journal of Culture, Tourism and Hospitality research. 2019;**13**(4):458-472. DOI: 10.1108/IJCTHR-03-2019-0044

[15] Stors N, Kagermeier A. Share economy in metropolitan tourism: The role of authenricity. In: Tózsa I, Zátori A, editors. Metropolitan Tourism Experience Development. Budapest: Department of Economic Geography and Furture Studies. Corvinius University of Budapest; 2015. pp. 90-103

[16] Sohrabi C, Alsafi Z, O'Neil N, Khan M, Kerwan A, Al-Jabir A, et al. World Health Organization declares global emergency: A review of the 2019 novel coronavirus (COVID-19). International Journal of Surgery. 2020;**76**:71-76. DOI: 10.1016/j.ijsu.2020.02.034

[17] Kuo H, Chi-Chung C, Wei-Chun T, Lan-Fen J, Bing-Wen H. Assessing impacts of SARS and avian flu on international tourism demand to Asia. Tourism Management. 2008;**29**(5):917-928. DOI: 10.1016/j.tourman.2007.10.006

[18] Henderson J. Corporate social responsibility and tourism: Hotel companies in Phuket, Thailand, after the Indian Ocean tsunami. Hospital Management. 2007;**26**:228-239. DOI: 10.1016/j.ijhm.2006.02.001

[19] World Travel and Tourism Council (W.T.T.C.). Special SARS Analysis: Impact of Travel and Tourism (Hong Kong, China, Singapore and Vietnam Reports). London: W.T.T.C; 2003

[20] Gursoy D, Chi C. Effects of COVID-19 pandemic on hospitality industry: Review of the current situations and a research agenda. Journal of Hospitality Marketing & Management. 2020;**29**(5):527-529. DOI: 10.1080/19368623.2020.1788231

[21] Hospitality and Catering News. Coronavirus: 80% Of Chinese hotel rooms empty. 2020. Available from: https://www.hospitalityandcateringnews.com/2020/02/coronavirus-80-chinese-hotel-roomsempty/ [Accessed: January 9, 2021]

[22] Research Institute for Tourism. Έρευνα Συγκυρίας & Επιπτώσεις από τον COVID-19 στην Ελληνική Ξενοδοχία [Situation Research & Impacts of COVID-19 in the Greek Hotels]. Athens: Hellenic Chamber of Hotels; 2020

[23] Pateli A, Kourouthanassis P, Nikopoulou M, Chasapi G. Digital transformation for resilient tourism: Evidence from the Greek hospitality industry. In: Christou E, Fotiadis A, editors. Reviving Tourism, in the Post-Pandemic Era. Thessaloniki: International Hellenic University; 2022. pp. 69-80. DOI: 10.5281/zenodo.6428590

[24] Sociology WT. Work and Industry. 4th ed. London: Routledge; 2003. p. 328. DOI: 10.4324/9780203103043

[25] Kvale S. Interviews, an introduction to qualitative research interviewing. American Journal of Evaluation. 1996;**19**(2):267-270. DOI: 10.1016/S1098-2140(99)80208-2

[26] Hallebone E, Priest J. Business and Management Research: Paradigms and Practices. Hampshire: Palgrave Macmillan; 2009. p. 252. DOI: 10.1007/978-1-137-11078-7

[27] Kokolakis G, Spiliotis I. Εισαγωγή στη Θεωρία Πιθανοτήτων και Στατιστική [Introduction to Probability Theory and Statistics]. Athens: Simeon; 1999. p. 368

[28] Gratton C, Jones I. Research Methods for Sport Studies. London: Routledge; 2003. p. 304. DOI: 10.4324/9780203380567

[29] Cohen L, Manion L. Research Methods in Education. 6th ed. London: Routledge; 2007. p. 656. DOI: 10.4324/9780203029053

Chapter 15

The Ongoing Impacts of the COVID-19 Pandemic on the Traditional Art Industry in North-Central Ethiopia

Tesfaye Fentaw Nigatu, Zelalem Getnet Ambaw and Fentanew Alelegn Masrie

Abstract

The main aim of this study was to assess the impact of the COVID-19 pandemic on traditional art and craftworks in North-Central Ethiopia. The study was conducted using a descriptive study design. A total of 72 samples were selected for a questionnaire survey, and 16 informants were contacted for interview purposes. The study found that 79% of the art businesses were closed, creating more visible and systematic challenges. The pandemic has created challenges for artists to change their work from crafts to mask production, traditional art and crafts pedagogy, and women's labor forces. Additionally, due to the COVID-19 pandemic, 80.56, 81.94, and 81.94% of traditional artists and craftsmen's physical mobility were paused, the artistic freedom and knowledge transfer activities were denied, and the ability to work their artistic activities was reduced, respectively. On the other hand, because of the COVID-19 pandemic, 77.78% of traditional artists and craftsmen were worried about being unemployed and 76.39% were in disturbing situations due to the coronavirus. Furthermore, 61.11% of respondents did not feel motivated or optimistic about the future of their work as a result of corona.

Keywords: artists, COVID-19, traditional art industry, north-Central Ethiopia, impact

1. Introduction

Art industry is one of the tertiary economic sectors and the backbone of many countries' economies. Most of the time art sectors are expressed in other similar terms like creative industry or entertainment sector or cultural industry or creative economy. Many researchers used these terms interchangeably [1–3]. The art industries are an industry that comprises a range of human activities in creative visual, auditory, or performing artifacts, expressing conceptual ideas or technical skills. The Britain Department for Culture, Media and Sport (DCMS) define the term creative industry as *"those industries which have their origin in individual creativity, skill and talent and which have a potential for wealth and job creation through the generation and exploitation of intellectual property*" [4, 5]. Creative industries/art industries are any economic activities that are concerned with the generation and exploitation of knowledge and information. These economic activities majorly incorporate craft, design, fashion, film, music, performing arts, photo and video graphs, museums, galleries, libraries, architecture, advertising, visual arts, TV and radio, computer games, and many others. These arts/creative industries have been seen to become increasingly important in economic well-being and we call them as the "twenty-first century industry" [6].

In addition to the economic benefits, the development of art and creative industries bring forth cultural value. These industries promote social inclusion because the nature of their activities links social groups together and strengthen cohesion among peoples. They also play a key role in the educational system, providing inputs that facilitate social and cultural education. Cultural and creative or art industries have indirect effects on other development areas due to the transverse nature of the industry [7]. In general, art industries contributed to the creation of jobs; increase of local economic incomes; enhancement of tourism management; reinvigoration of urban areas; retention and attraction of talent and investment; and social integration and rationalization of infrastructure. A dynamic creative sector will also help to empower women, young population, rural groups, and even isolated communities by using their national cultural heritage, and their own knowledge and creativity [8, 9].

Crisis was heard everywhere at the moment. According to International Monetary Fund (IMF), 2020 is likely to be the worst year for the global economy due to the sudden outbreak of coronavirus (COVID-19: CO implies Corona, VI implies Virus, D implies disease, whereas 19 indicates the year of its outbreak (2019)). The art industries were also the major sector affected by both natural and manmade calamities. COVID-19 is one of the twenty-first century global pandemic that greatly paralyzed the world industry; art sector is one of its target. After the first infections in China at the end of 2019, the coronavirus disease (COVID-19) has continued to spread across the world. No continent has been able to escape this virus since the declaration as a pandemic by the World Health Organization (WHO) on March 11, 2020; COVID-19 has become a global emergency, given its impact on the entire world population and the economy [7, 10, 11].

The arts sector serves artistic and educational missions through myriad occupations filled by individuals who work in full time, part time, hourly, seasonal, and freelance capacities. Based on the American Bureau of Economic Analysis reported in March 2020, the arts and culture workforce contributed $877.8 billion, or 4.5% to the nation's gross domestic product (GDP) in 2017. The arts sector is an economic engine by providing employment opportunities for more than 5 million workers in the United States [12].

The effect of COVID-19 on economies around the world has been unprecedented. While evidence of this impact is only beginning to emerge, it is clear that the economic damage has been particularly severe for Small and Growing Businesses (SGBs) in emerging markets. Just as small and growing businesses, art and craft sectors of the economy are majorly tested in their "sustainability" [13]. A multitude of shows, performances, festivals, film shoots, book fairs, and other cultural events was canceled, and many cultural spaces, bookstores, cinemas, and museums closed their doors because of the outbreak of COVID-19. Therefore, the pandemic creates a lasting impact on the life and livelihoods of the employees, organizers as well as the art industry in general. United Nation World Tourism Organization (UNWTO) affirmed that the outbreak of COVID-19 as a world pandemic affects the fundamental rights of access to culture, the social rights of artists and creative professionals, and the protection of a diversity of cultural expressions [14, 15]. Generally, the overall impacts of the pandemic on the art and craft sector are just becoming apparent and with the overall impacts still unknown.

United Nation World Tourism Organization (UNWTO) called on governments, intergovernmental organizations, civil societies, and educational institutions to mobilize their resources to study the multidimensional impacts of the pandemic on the art industry and recommend implementation and policy direction. The government of Ethiopia also encourages research-based intervention on the impacts of COVID-19 pandemic in all sectors of the economy as well as its impact on the social and cultural life of the community. Ending surveys studies are conducted in different parts of the country on the issue of health matter.

It is apparent that there are huge number of traditional night club, traditional souvenir shops, producers, and suppliers of traditional crafts in Amhara region, particularly in Lalibela, Dessie, and Woldia as well. As repeatedly disclosed in media, all these sectors are closed their doors and both art and artists become victims of COVID-19 pandemic. Above all, despite the fact that the region is known for such type of art, currently, they could not find due attention by researchers and government as well. We have been conducted and affirmed in our preliminary assessment that both the magnitude of the current impact and its long last impact are not yet studied so far. Therefore, this study was developed in order to fill the gaps.

2. Objectives of the study

- To describe the continuing impact of COVID-19 pandemic on traditional art and craft sector supply and mobilization of input and output resources in North-Central Ethiopia.

- To analyze the impacts of COVID-19 pandemic on traditional artists and craftsmen's life in the study area.

3. Methodology of the study

3.1 Approach and design

To achieve this subtheme, the researchers were guided by using of mixed-method study design. It encompasses both qualitative and quantitative approaches. Using of mixed research approach is very significant to gain the advantage of or the strength of both qualitative and quantitative research that can provide the best understanding of the issues under study [16]. Furthermore, a mixed approach is important to reconcile the limitations with the representativeness of the sample of the qualitative approach. In this regard, this study employed both research methods, so as to increase the validity of the quantitative data and truth worthiness of qualitative data. In research concept, research design is an idea that expresses the procedure for collecting, analyzing, interpreting, and reporting the data. Thus, this subtheme followed a descriptive cross-sectional survey research design, where the data collection process is conducted at a point in time (one-shot data from the sampled study population). This cross-sectional descriptive design was used to examine the issues at hand by using a concurrent triangulation approach because both quantitative and qualitative data were gathered simultaneously [17]. In order to assess subtheme one (the impacts of COVID-19 on the arts industry) the researchers utilized both qualitative and quantitative data.

3.2 Data sources, method of data collection, and sample frame

The researchers collected both primary and secondary data. The primary data were collected by using a questionnaire, interview and focus group discussion (FGD), and personal observation techniques. The sample frame for this subtheme are traditional art performers-from actors to musicians, crafters, souvenir shoppers, culture and tourism officers, suppliers of traditional art and crafts sectors, and marketers in traditional art and craft. The personal observation was conducted in traditional arts center and crafts areas.

4. Sampling techniques

The samples were selected by using a purposive sampling technique. The samples were selected from the major towns/cities of the region (Dessie, Lalibela, Woldia, Haik, and Kombolecha). We contact 118 people for questionnaire and 16 people for interview to accomplish this subtheme. A group of people gathered and discussed the issues under investigation. We have five FGDs (one in each town/city). The researchers were taken samples in the major cities by considering the large number of art and crafts sector personnel in the cities. Therefore, we developed the following sample proportion in each city and town (**Table 1**).

5. Data analysis and interpretation

5.1 The impacts of COVID-19 pandemic on traditional art sectors

5.1.1 Demographic characteristics of the respondent artist and craftsman

As shown in the demographic chart above, 59.7% of the respondents were male and while 40.3% were female. The respondents were young because almost 51.4% and 37.5% of the respondents were found in the age range of 18–28 years and 29–39 years, respectively. Regarding educational level, 52.8% and 36.1% were secondary education

No.	City/Town	Number of samples	Remark
1	Dessie	38	
2	Lalibela	43	
3	Kombolecha	20	
4	Woldia	11	
5	Haik	6	

Table 1.
Sample proportion table of traditional artists and craftsmen.

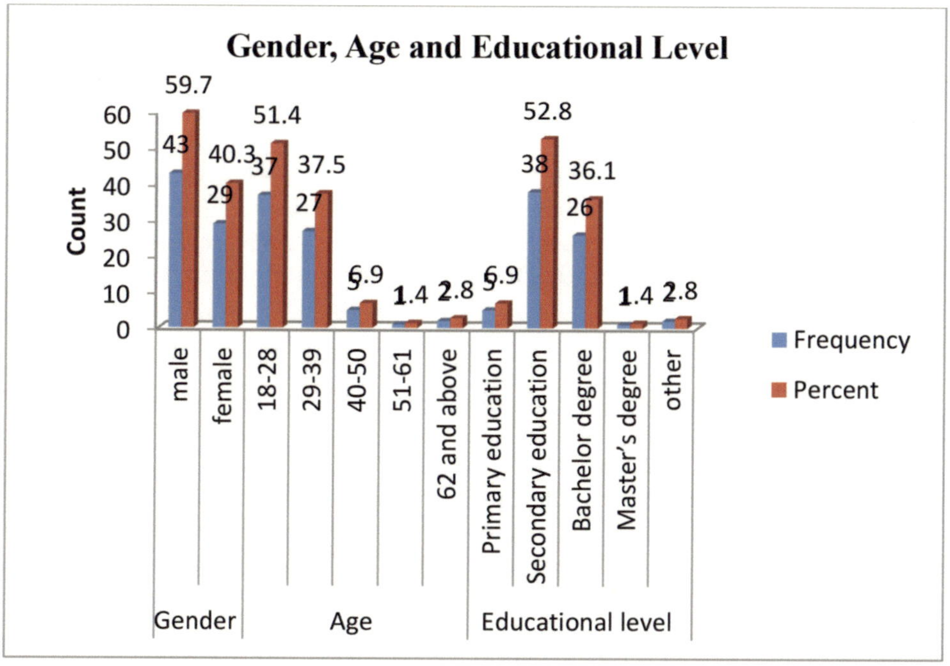

Figure 1.
Demographic characteristics of the respondent artist and craftsman. Source: Researcher's survey, 2022.

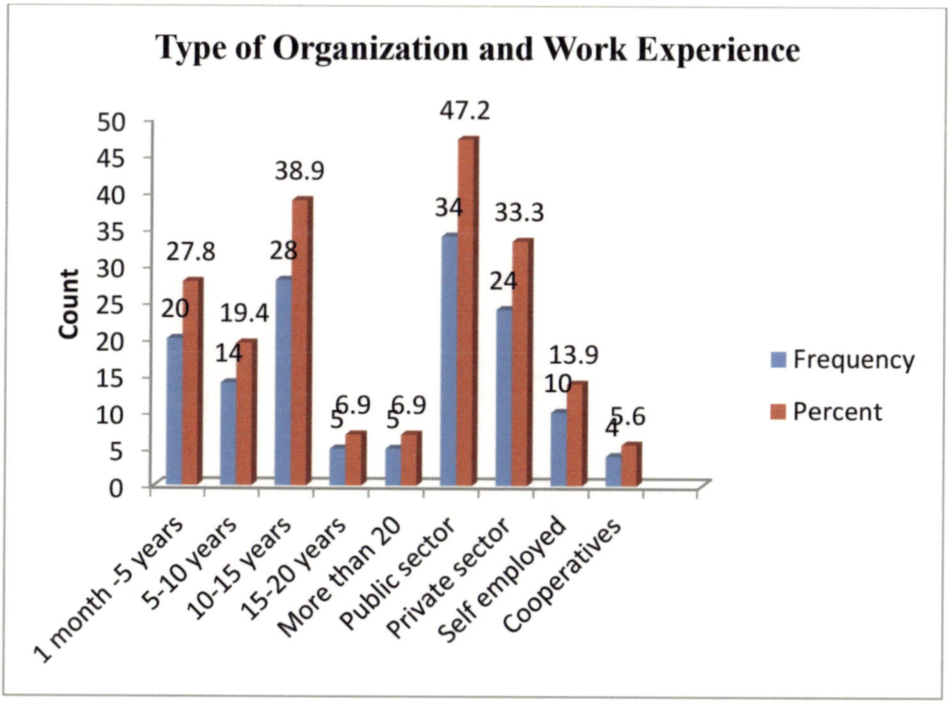

Figure 2.
Type of organization and work experience of the respondent artist and craftsman. Source: Researcher's survey, 2022.

and bachelor's degree completed, respectively. These figures indicated that the majority of the respondents were male, young, and have some educational level with secondary and bachelor (**Figure 1**).

As shown in **Figure 2** above, 38.9% of the respondents experienced 10–15 years in the sector of the art industry. About 27.8% of the respondents had working experience from 1 month to 5 years, while 13.8% had more than 15 years of working experience. Regarding the type of organization they worked in 47.2% worked in the public sector, while 33.3% worked in the private sectors. Only 5.6% and 13.9% worked in cooperatives and self-employed, respectively. According to this survey information, more than half of the respondents worked more than 10 years in the sector and experienced (**Figure 3**).

5.1.2 The multidimensional impacts of the pandemic on artists and traditional craftsmen's life

Besides the general impacts of the industry, it also has a great impact on the lives of traditional artists and craftsmen. The survey result indicated in the following chart also indicates the status of the art business during the lockdown. Almost 79.2% of the art business was closed in the study area while only 20.8% of art businesses are opened (**Figure 4**).

The interview with traditional artists and craftsmen in our study area agreed on the fact that COVID-19 has created more visible and systematic challenges, especially in their economic instability due to loss of income, irregular working conditions, and more time business shutdown. This problem is also exacerbated because of the gaps in our cultural policy. One of our interviewees indicated that Ethiopian cultural policy is very weak to handle such global catastrophe and support artistic workers and crafters.

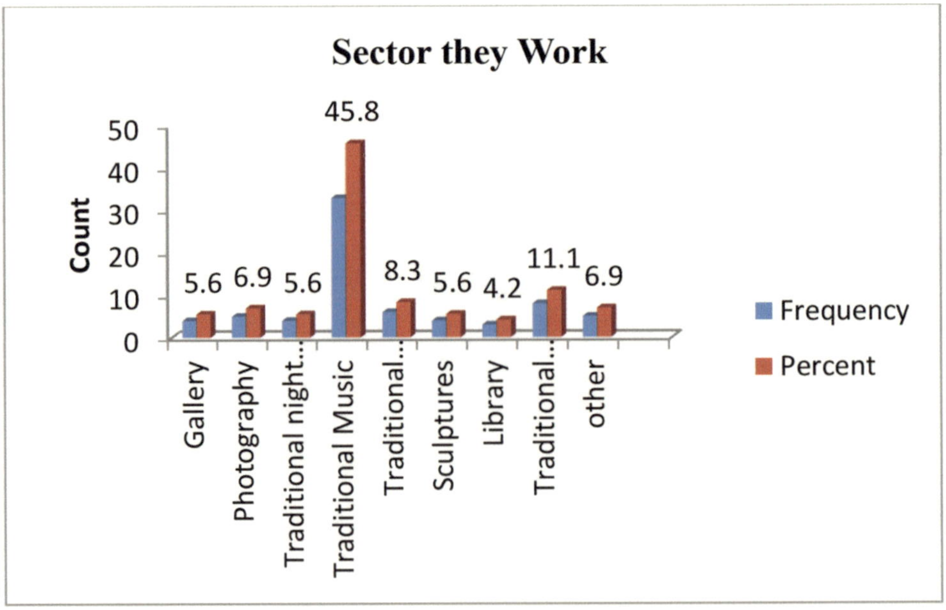

Figure 3.
Sector they work. Source: Researcher's survey, 2022.

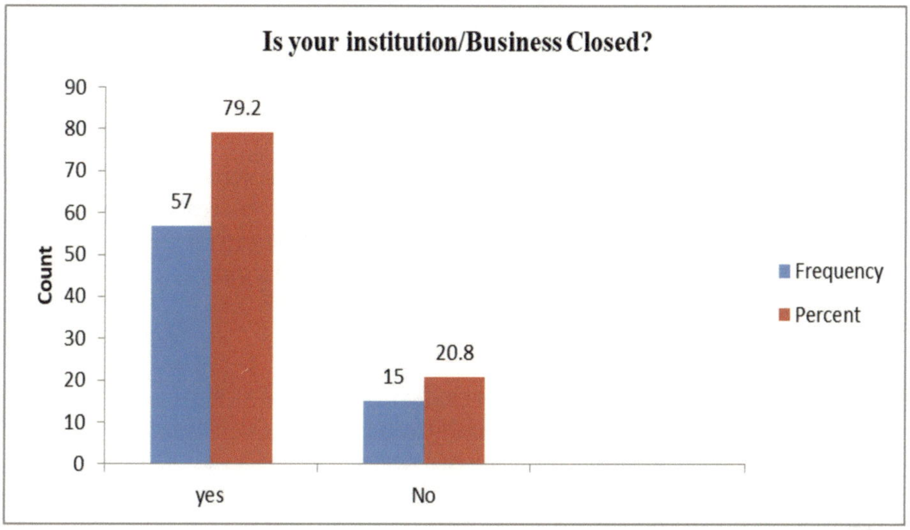

Figure 4.
The status of the art business. Source: Researcher's survey, 2022.

Due to the weakness of the cultural policy, many artists and craftsmen left alone with a low priority of government support and "uncertainty incentive."

The major impact of COVID-19 on the artists, artistic lives, and craftsmen during the pandemic was, it totally break their social connection. One of our interviewees describes these challenges as "due to COVID pandemic my social connections with audiences, peeks, social networks, business partners...therefore, I am not able to create artworks, feeling of lonely, able to perform art work and unable to plan the future." On the other hand, many respondents said that COVID-19 greatly affected their life and created a plethora of issues, from possible cancelations of orders to a lack of awareness about the disease. They affirmed that their business and practices of the craftsman by the disease because many sales opportunities have been postponed and the remaining canceled. As one of our traditional cloth designers told us "I am always worried about the employees working with me all the workers in my small garment the lives depends on it; we expect more work in the coming July and September 202/22, however due to the economic slowdown our customers did not make any order for the new year festivity." The COVID-19 crisis has had a multidimensional impact on all aspects of our lives around the world, but some sectors have been hardly affected. Among these sectors, traditional art sector is one of the sectors greatly affected. Many exhibitions, festivals, traditional art concerts, events, meetings with artists, and meeting that need the involvement of artists and musicians are also canceled. In this section, we present some of the short- and long-term impacts of COVID-19 pandemic on the art sector in the study area from FGD and interviews of the respondents. In general, the following multidimensional impacts are observed.

a. Event cancelation: Museums, traditional bars and restaurants, and theatrical activities were closed for more than 6 months. For example, Dessie museum closed almost a year. Due to these closer many events were canceled and no travelers visited the areas.

b. Fundraising activities slowed: Due to coronavirus, many sponsors and donors of traditional artists, musicians, and performing artists slowed and some canceled. Therefore, many artistic sectors lost their sponsorship and donation revenues for their service.

c. Revenue lost: Many of the economic sectors of the area were greatly affected by COVID-19 pandemic. Many of the tradition art sectors, restaurants, hotels, festivals, and tourism-related businesses completely closed in the study area for more than 3 months. Some were also partially closed for almost half of the year. These also created a huge economic loss in the study area, especially in the areas of Lalibela and Dessie.

d. The capacity of the industry was reduced: Due to the reduced revenue, the potential loss of sponsors and donors the institution lost its institutional knowledge and a reduction of capacity to handle programs and services.

e. Many art and craft workers changed from crafts to mask production: The great challenge the art sector faced is the reopening and practicing of the art business cannot happen overnight, like other business because the art sector business requires an up-to-date skills and psychological makeup and long-term awareness creation among the audiences and customers.

f. Its impact on the traditional art and crafts pedagogy: For almost a year many traditional arts and crafts educational systems ceased in the study area. In our country in general and in the study area in particular digital education *via* video and practical online platform was impractical. Therefore, many technical and vocational training in the field of art and crafts work stopped totality.

g. Its greater impact on women labor forces: COVID-19 created a devastating impact on the women labor force globally, especially those works more dominated by women as "frontline workers" such as art and hotel sector. According to World Health Organization report from 104 countries, 70% of workers were exposed to health risks. Among these, 43% were women because of the nature of the work they engaged (WTO, 2020). Due to the fact that women are over-represented in the industries most affected by the pandemic, such as recreation and entertainment sector, craft manufacturing, education, and other service sectors, they are highly affected by the pandemic. One of the female respondents expressed her feeling on the impacts of COVID-19 on the labor force of women "for the last 11 years I working as traditional cloth designer (handloom) in addition to the informal household works but now I am out of work and always working as a meal planners in the house...when I calculate the working hours as compared to the previous I spend additional 2 hours because all the families see my hands even we consume more food when we are in home... even I sometimes worried about we may affected by the food shortage." In overall way, the pandemic affected the art industry greatly. The survey also attested that more than 80% of the art and traditional businesses were affected. It is shown in the chart below (**Figure 5**).

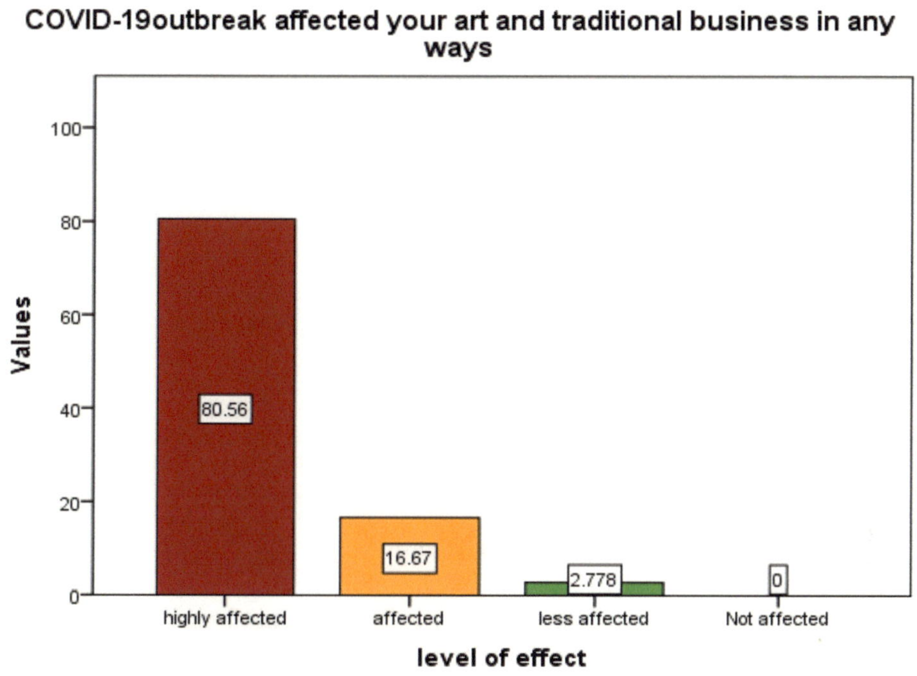

Figure 5.
The overall effect of the pandemic. Source: Researcher's survey, 2022.

5.1.3 Working conditions of traditional artists and crafts during the pandemic

The outbreak of COVID-19 pandemic created a complex working environment and lives among artists and craftsmen in the study area. The first and foremost impact of COVID-19 pandemic is the behavioral change of art audiences. One of our informants said that "I did not believed that demands will be restored as a previous level" (performing art organizer, September, 2021). FGD discussants added that due to COVID-19 pandemic the behavior and attitude of the audiences of art and performing art changed for a prolonged time. The pandemic affects the spending power, visitation behavior due to the economic recession, and lower socioeconomic status of the participants due to COVID-19 pandemic. Other studies conducted by Shugoll research survey found that 29% of the respondents believed that they will wait 6 months and more to return to the theater again.

Many traditional artists and craftsmen are affected by the outbreak of COVID-19 pandemic. Many of nightclubs, music performances, and contests [the workplaces of traditional artists and craftsmen] closed for the long period of time (almost a year). This has created a complicated life and loss of sources of revenue for these portions of the art workers (**Figure 6**).

As indicated in the above chart the working conditions of traditional artists and craftsmen during the pandemic are shown. From this survey result, 80.56% of the respondents said "Yes" for the question the pandemic pause your physical mobility? And, 79.17% said "Yes" for the question the pandemic stop the grants and loans from the GOs and NGOs? 81.94% of the survey respondents affirmed that due to coronavirus, their artistic freedom and knowledge transfer activities were denied. The ability

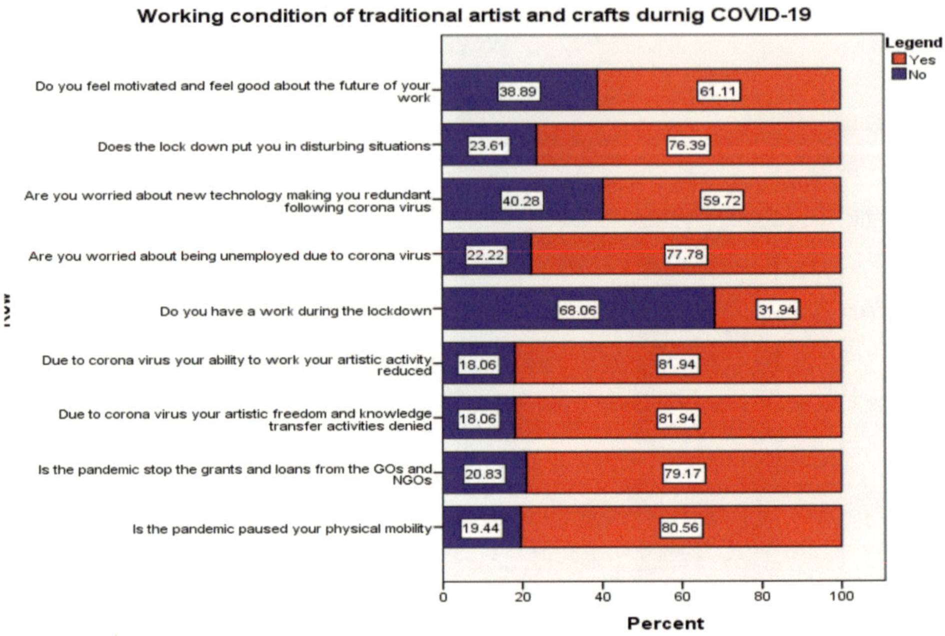

Figure 6.
Working conditions of traditional artists and craftsmen. Source: Researcher's survey, 2022.

to work in artistic activity was reduced by 81.94% among the artists and traditional craftsmen during the corona. Besides, 77.789% of the traditional artists and craftsmen were developing a feeling of worried about being unemployed due to coronavirus and 59.72% of the respondents were worried about new technology making them redundant following the coronavirus. On the other side, 76.39% of the surveyed respondents develop a feeling of disturbing situations during the lockdown. However, the majority (61.11%) were feeling motivated and feeling good about the future of their work and 68.06% were having work during the lockdown.

5.1.4 The impact of COVID-19 pandemic on traditional art and craft sector suppliers and on the mobilization of input and output resources

The outbreak of coronavirus created a series of blows on traditional arts and craft sectors' raw material mobilization. The lockdown affected every production and distribution of raw material supplies. The transportation system and the attitude change of the suppliers created a bottleneck for the input mobilization. When we say attitudinal problem one of our interviewees of traditional arts expressed that "our suppliers and customers assumed and think that traditional cloth and craftwork considered as a luxury category of products... no one want to buy such products in this difficult time. Therefore, we totally stopped our work for a long period of time ... to tell you truly almost a year we are out of business due to coronavirus. Really it is a difficult time to survive..." The survey result also strengths the FGD discussants and interviewed participants. As shown in **Table 2**, 83.3% of the art industry suppliers did not provide input for the art industry principals during COVID-19 pandemic era. The same is true in the mobilization of input resources because 88.9% of the mobilization of input and outputs did not properly hold during the lockdown. However, even with the problems exist in the mobilization of the input and output resources, 73.6% had a

Questions	Yes (%)	No (%)
Does your suppliers did provide input for your business during corona	16.7	83.3
Is the mobilization of input and outputs properly held during the lockdown	11.1	88.9
Do you have a plan to invest more in your business in the next 5/10 years	73.6	26.4

Source: Researcher's survey, 2022.

Table 2.
Supply and mobilization of input and output resources.

plan to invest more in their business in the next 5/10 years, while 26.4% did not have the plan to invest more in their business in the next 5/10 years.

5.1.5 The implications and mechanisms used on the future development of traditional arts and crafts industry

The outbreak of COVID-19 pandemic has many implications for the art business in the study area. This was also assessed by a survey in the traditional art business personnel and shown in the above chart (**Figure 7**). As indicated above due to the outbreak of coronavirus, 69.44% of art business owners and runners could not meet the cost of input for their business. Additionally, 72.22% of the respondent art business owners could not make utility payments like water and electricity, and 77.78% of respondents were forced to borrow from their families and friends to run their businesses due to the impact of COVID-19 pandemic. About 59.72% of art business owners faced a shortage of cash to pay the salary of their employees because of coronavirus. The most visible impact of the pandemic is on the future scale-up of the art business because the majority (93.08%) of the surveyed respondents affirmed that could not scale up the operation of the business. On the other hand, 94.44% of the respondent art business operators could not produce or deliver service as before because of COVID-19 pandemic in the study area. The outbreak of the pandemic also had a great impact on the tax payment. The survey also affirmed that 59.72% and 66.67% of the respondents could not make tax payments in full and could not make tax payments on time due to COVID-19 pandemic, respectively (**Figure 8**).

The survey also assessed the skills and resources available to respond to the impact of the coronavirus, including future audiences behavior (54.17% important and 29.17% very important) innovation/digitalization (41.67% important and 44.44% very important), technological capacity (30.56% important and 58.33% very important), leadership support (40.28% important and 50% very important), training (51.39% important and 33.33% very important), staff skill (37.5% important and 52.78% very important), finding new funding sources (33.33% important and 59.94% very important), and response in responding to the impact of COVID-19 pandemic. Organizational restructuring, staff restructuring, introducing new programs and partners, and managing staff safety and security were also important to respond to the impact of coronavirus in the future of art business in the study area (**Figure 9**).

More than half percent (53.2)% of the art business operators needed more than 1 year to return back to normal business and life in the study area while the remaining needed less than 1 year to return back to their normal business (for more see chart 9 above).

As shown in **Figure 10**, the majority (44.4% and 34.7%) of the respondents needed loans with the low-interest rate and provision of subsidies from the

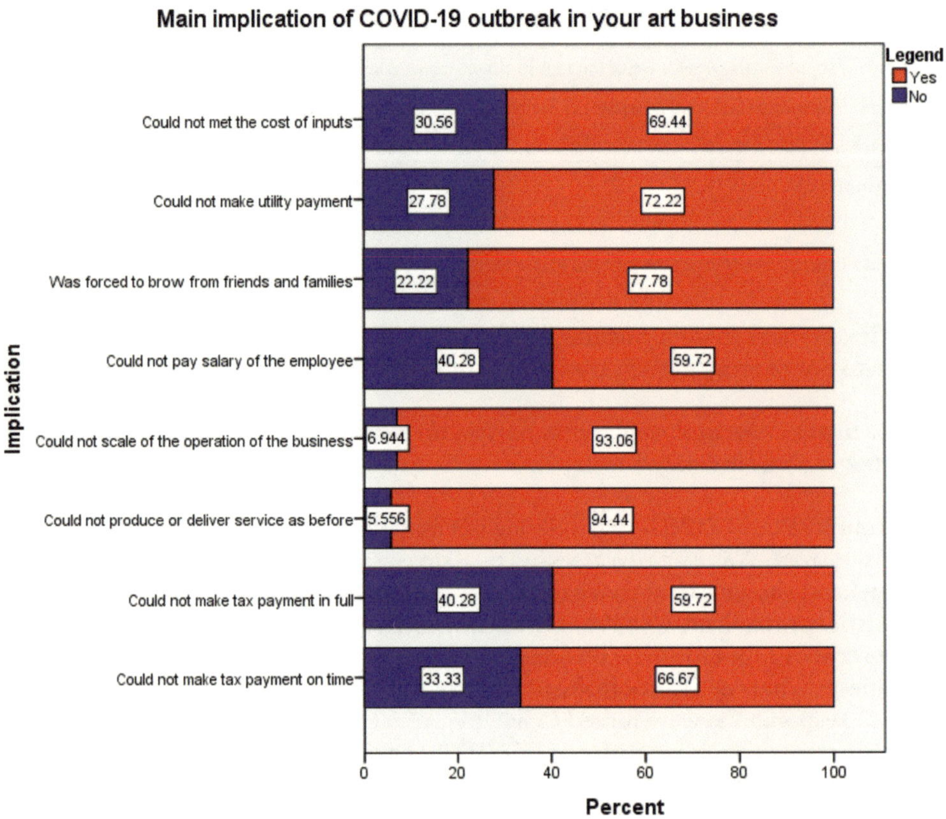

Figure 7.
The implications of COVID-19 on the business. Source: Researcher's survey, 2022.

government, respectively. Very few (1.4%) needed subsidies in the tax form from the government. Some others (19.4%) needed to allow running their business operation without any prohibition.

On the other part, 30.6% and 29.2% of the respondents' traditional artists and craftsmen prefer shut down their businesses completely and shut down their businesses partially, respectively. Only 15.3% of the respondents continue to run their business as a short-term method as a survival mechanism and some others (13.9%) wanted to change their business line and also some others (9.7%) apply for loan used as a mechanism during the lockdown (**Figure 11**).

6. Conclusion and recommendations

6.1 Conclusion

The study concludes that 79% of the art businesses were closed. Due to the COVID-19 pandemic traditional artists and craftsmen's physical mobility was paused, the artistic freedom and knowledge transfer activities were denied, and the ability to work their artistic activities was reduced. On the other hand, because of the COVID-19 pandemic, 77.78% of traditional artists and craftsmen were worried about being unemployed, 76.39% were in disturbing situations, and 61.11% did not feel motivated

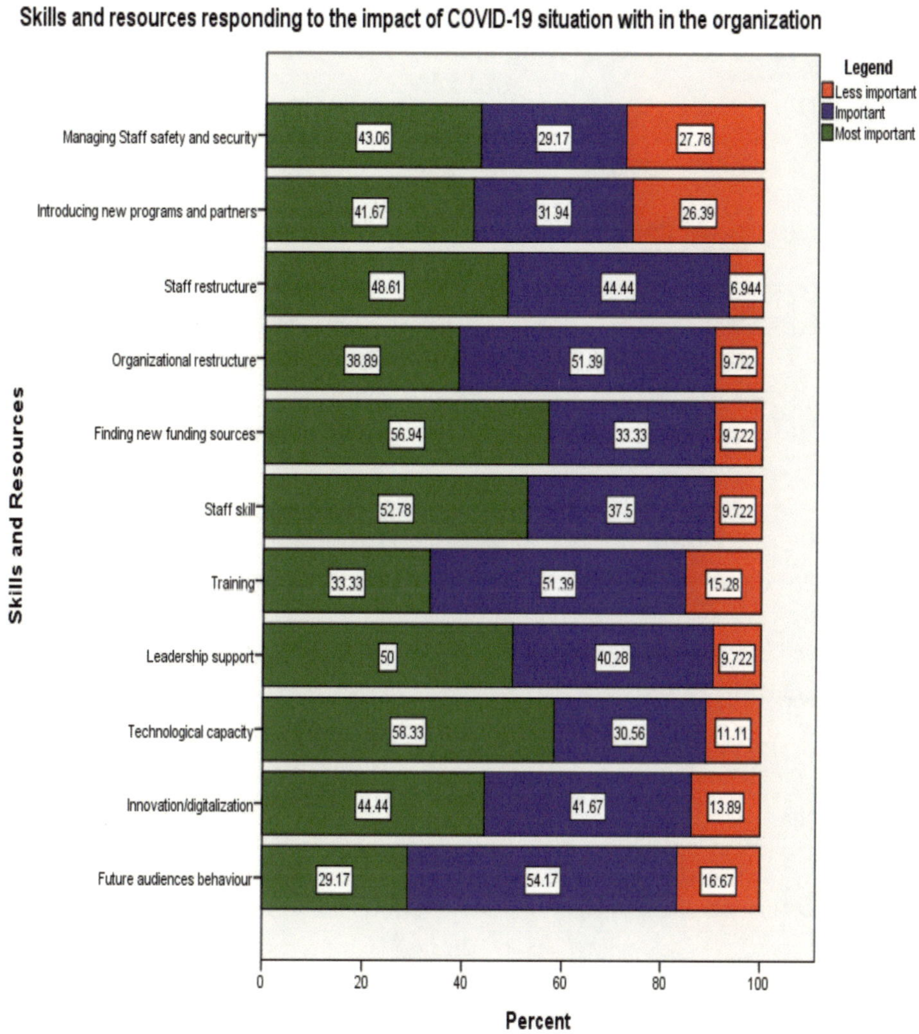

Figure 8.
Skills and resources responding to COVID-19 pandemic. Source: Researcher's survey, 2022.

or optimistic about the future of their work as a result of corona. Due to the coronavirus, 83.3% of art business suppliers did provide input, while 88.9% of the mobilization of inputs and outputs did not properly hold during the lockdown. The outbreak of the COVID-19 pandemic has many implications for the future development of the art industry in the study area. The study also assessed the skills and resources available to respond to the impact of the coronavirus, including innovation/digitalization, leadership support, training and staff skill, organizational restructuring, staff restructures introducing new programs and partners, and managing staff safety and security were identified.

6.2 Recommendations

Based on the finding of the study, the following possible recommendations were forwarded.

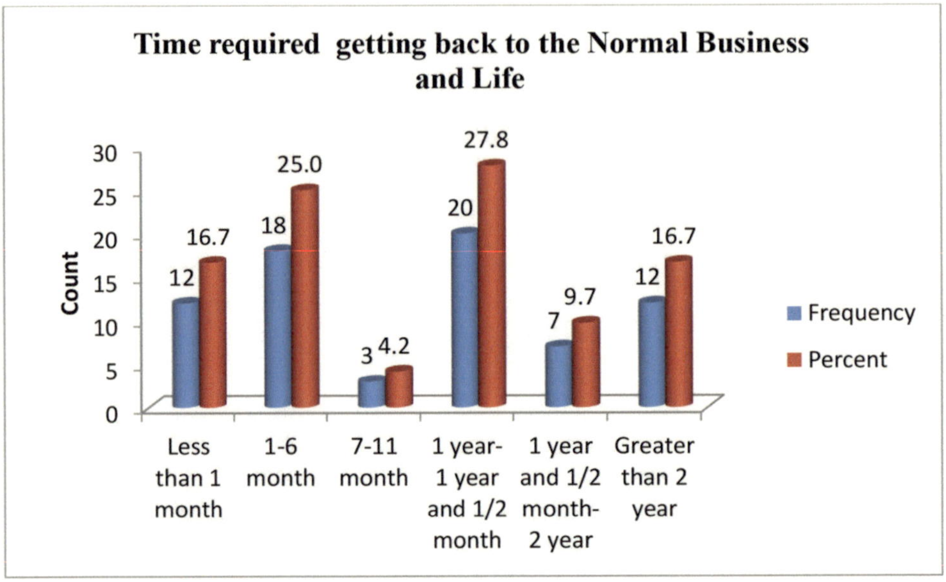

Figure 9.
Time required to get back to normal business. Source: Researcher's survey, 2022.

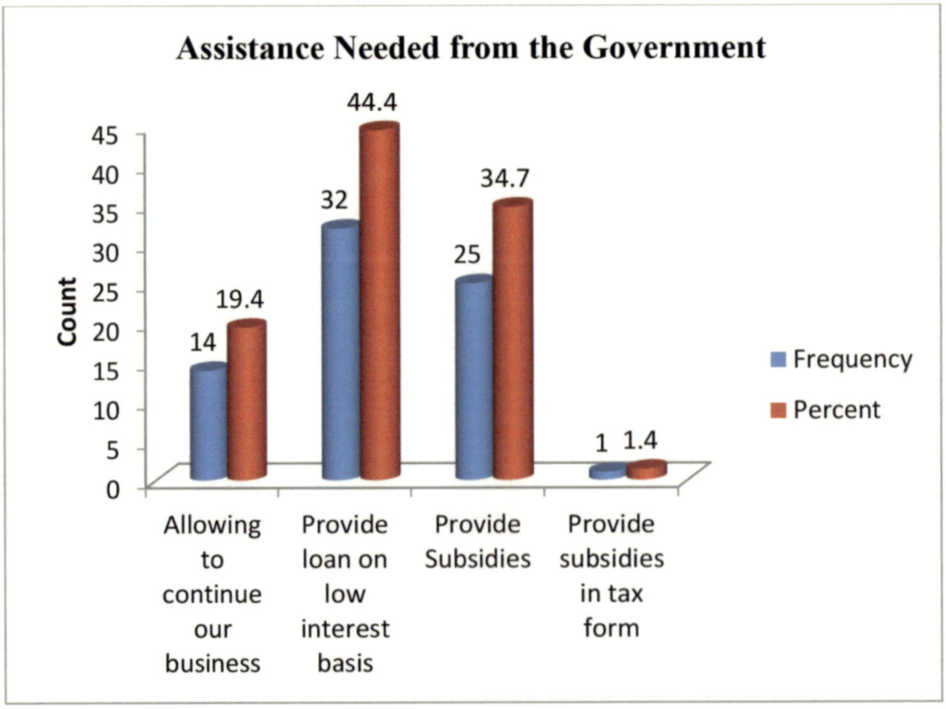

Figure 10.
Assistance needed from the government. Source: Researcher's survey, 2022.

- The national, regional, and local government offices should work on the improvement of working cbl aonditions of the traditional art and craftsmen because the working condition of the workers in the study area is in miserable

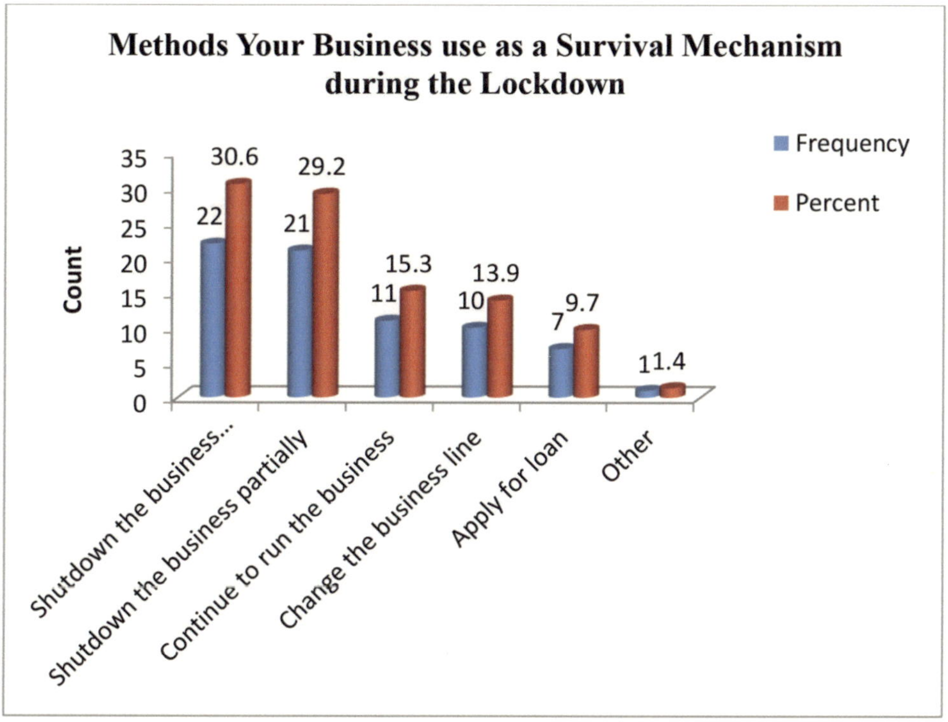

Figure 11.
Methods used as a survival mechanism during the lockdown. Source: Researcher's survey, 2022.

condition during the lockdown and it also has a psychological, social, and economic impact in their life.

- Empower the suppliers of the art industry *via* training and capacity-building works about the COVID-19 pandemic issues.
- Encouraging and supporting traditional artists and craftsmen/art business operators to use technology/digital media and other technology usages.
- Support the art business operators in training capacity-building work in leadership and management and finding new funding sources from the government bodies.
- Government officials and culture tourism offices at different level consult the art business operators in finding new programs and partners, staff skill development, staff restructuring, and organizational restructuring works.
- The government should provide loans with low-interest rate and provision of subsidies for art business operators for sustainability and continuation.

Acknowledgements

We acknowledged Wollo University Research and Community Service Vice president office, Research and Publication Directorate Office for supporting and financing the study. Sincerely, we would like to thank to all our respondents of this research work, who helped us to give their time during the interview process, and who share their ideas for the success of this study. We strongly acknowledged the anonymous reviewers for their valuable comments and suggestion for the entire development of this scientific paper.

Conflict of interest

The researchers declared that no conflict of interest in this particular study.

Author details

Tesfaye Fentaw Nigatu[1,2]*, Zelalem Getnet Ambaw[2] and Fentanew Alelegn Masrie[3]

1 Sunway University, Malaysia

2 Department of Tourism and Hotel Management, Wollo University, Dessie, Ethiopia

3 Department of Theatre, Wollo University, Dessie, Ethiopia

*Address all correspondence to: tesfaye.fentaw@wu.edu.et or 22118889@imail.sunway.edu.my

References

[1] Susan C. Beyond the self-expressive creative worker: An industry perspective on entertainment media. Theory, Culture & Society. 2010;**2008**(25):73. DOI: 10.1177/0263276408097797

[2] Caves R. Creative Industries: Contracts Between Art and Commerce. Cambridge, MA: Harvard University Press; 2000

[3] Cleary K. COVID-19: The pandemic's impact on mental health. In: News and Features. 2020. Available from: https://www.spotlightnsp.co.za/2020/05/04 covid-19- the-pandemics-impact-on-mental-health/

[4] Department for Culture, Media and Sport (DCMS). Creative Industry Mapping Document. 2nd ed. London, UK; 2001

[5] Shugoll Research. Coronavirus Theatre Survey. National Sample for Theatregoers. Quantitative Research Findings. Bethesda, MD. 2020

[6] Howkins J. The Creative Economy: How Peoples Make Money From Ideas. USA: Penguin Global; 2001

[7] United Nation Industrial Development and Organization. Music Industry's Contribution Towards Inclusive and Sustainable Development: The Case of Cuba, the Music Island. 2017

[8] National Bureau of Statistics of China. Impact of Covid19 in the First Two Months. Beijing, China. 2020. Available from: National Bureau of Statistics of China (stats.gov.cn)

[9] UNESCO. COVID-19 Educational Disruption and Response. 2020

[10] United Nation Word Tourism Organization. Culture and COVID-19: Impacts and Response Tracker. Issue 1. 2020

[11] UNWTO (United Nations World Tourism Organization). 2020: Worst Year in Tourism History With 1 Billion Fewer International Arrivals. (News Release. 28 January 2021). Madrid; 2021. Available from https://webunwto.s3.eu-west-1.amazonaws.com/s3fs-public/2021-01/210128-barometer-en.pdf?GaI1QTYG.Ky9LDZ2tlDKc.iRZkinJeuH

[12] Wyman O. To Recovery and Beyond: The Future of Travel and Tourism in the Wake of COVID-19. World Travel and Tourism Council; 2020. Available at https://wttc.org

[13] Arts Council. Employment and Economic Impact Assessment of Covid-19 on the Arts Sector in Ireland. Dublin, Ireland; 2020

[14] Arutz S. Thousands to Prey at Western Wall for End to COVID-19 Epidemic. Arutz Sheva, Israel National News; 2020. Available from http://www.israelnationalnews.com/News/News.aspx/276027

[15] UNWTO. COVID-19 and Vulnerable Groups: An Inclusive Response for Vulnerable Groups'. 2020. Available from: https://www.unwto.org/covid-19-inclusive-response-vulnerable-groups

[16] Veal AJ. Research Methods for Leisure & Tourism: A Practical Guide. 4th ed. England: Prentice Hall; 2011

[17] Creswell JW. Qualitative Inquiry and Research Design: Choosing Among Five Traditions. Thousand Oaks: Sage Publications; 1998

Chapter 16

The Long-Term Impact of COVID-19 on Inbound Tourism from China: Using 2020/2022 Web-Based Survey Data

Takeshi Kurihara, Kazuo Nishii, Yilin Sun and Yinan Dong

Abstract

This study discusses the long-term impact of the COVID-19 pandemic on inbound tourism from China, aiming to investigate its prospects during the post-pandemic period. After briefly reviewing trends concerning COVID-19 impact studies at home and abroad, basic results from two cross-sections of web-based data in 2020 and 2022 are introduced to identify how the pandemic impacted not only daily activity and travel patterns but also the intentions of visiting Japan in the post-pandemic period. Finally, we summarize the challenges that we should verify to support inbound tourism restoration policies.

Keywords: COVID-19's long-term impact, inbound tourism from China, post-pandemic tourism policy, visit intention, cross-sectional survey data

1. Introduction

In Japan, since the state of emergency due to the COVID-19 pandemic was declared on April 16, 2020, we have experienced several pandemic waves due to the emergence of new types of coronaviruses. The eighth wave of the pandemic caused by the XBB.1.5 subvariant in November 2022 is heading to a conclusion by mid-March 2023 [1]. The impact of the COVID-19 pandemic over a period of more than three years on socioeconomic activities is evidently significant enough to trigger not only changes in the attitude toward the risk of being infected in daily activities and travels in the short term, but also those in both their intentions and their real behavior patterns and lifestyles in the long term, that is, in the post-pandemic period [2]. Many analytical studies are attempting to quantitatively understand the actual conditions of changes in daily activity patterns during the pandemic period.

In the case of overseas COVID-19 impact studies, data collection and surveys on people's attitudes toward the risk of infection and changes in their activity and travel behaviors have been successively conducted since the WHO declared the COVID-19 pandemic on March 11, 2020. Following this data collection, many empirical results from studies on the relatively short-term impact of the COVID-19 pandemic were reported at several workshops and conferences, such as the ERSA Workshop [3] and TRB Annual Meeting [4]. These studies were mainly concerned with behavioral changes, comparing data from the pre-pandemic period with those after the first pandemic peak (i.e., from June to August 2020).

Against this background, this study aims to examine the prospects for a post-pandemic recovery in tourism demand for visiting Japan from China and to clearly identify the challenges of achieving sustainable and resilient inbound tourism destinations in Japan. While COVID-19's long-term impact is of particular concern in our study, let us explain that such a dynamic process would be founded on the hypothesis that both the repeated appearance of a new type of coronavirus and the related preventative policies trigger changes in activity and travel patterns, which would be partially established throughout the pandemic period with multiple waves, and become common as a lifestyle called the New Normal in the post-pandemic period. It is also hypothesized that such a shift to the New Normal should be evaluated as an inclusive and long-term impact, taking into consideration the effects of socioeconomic circumstances such as an aging society with fewer children, digital communication innovation (called DX [Digital Transformation]), and a low-carbon society with the GX [Green Transformation]). To examine the long-term impact on tourism demand for visitors to Japan (abbreviated as VTJ) from China, our study quantitatively identifies relationships among various kinds of factors related to COVID-19 impacts over time. Specifically, we empirically analyze both changes in attitudes toward COVID-19 risks and those in daily activity and travel behaviors in response to governance measures over the past three years. Based on the results of the 2020/2022 web-based survey data analysis, we intend to specify the factors and preconditions that determine the long-term impact of COVID-19 on VTJ demand from China and summarize the challenges toward sustainable tourism destination management in the post-pandemic period.

Since 2019, we have addressed the demand for integrated mobility services, targeting VTJ tourists from China (see refs. [5, 6]). This preceding study motivated us to our current COVID-19 impact analysis. As the COVID-19 pandemic occurred in China at the end of December 2019, we decided to add COVID-19-related items to the 2020 web-based survey and judged the timing of the survey implementation considering the actual conditions of the pandemic in the three targeted cities (Beijing, Zhejiang province, and Jiangsu province). The 2020 web-based survey was conducted at the end of November 2020, when the first wave of the pandemic was settled and these cities were released from severe behavioral restrictions, such as lockdowns in the district [7, 8]. In this study, as mentioned in the following section, the second web-based survey conducted in mid-April 2022 targeted seven cities, adding four cities (Shanghai, Guangdong, Shandong, and Liaoning provinces) to the three previous ones [8]. The 2022 data used for our COVID-19 impact analysis were compared with those of the 2020 web-based survey. Therefore, this study intends to explore the actual changes in not only people's daily activity and mobility patterns and their intentions of domestic and overseas tourism travel at two points in time, but also their intentions of VTJ travel under the condition that overseas travel restriction measures such as entry/exit visa restriction and quarantine would be lifted due to the reduction in the risk of COVID-19 infection.

2. Literature review

2.1 Review of COVID-19 impact studies in Japan

Various COVID-19 studies in Japan have been triggered by the "arrival of the corona disaster" in which the government made an emergency declaration for the first pandemic and decided to promote preventative measures, such as behavior restriction, self-restraint from going out and the three Cs, "Closed spaces, Crowded places, and Close contact," in April 2020 [9].

They launched a data collection and survey in three sectors, industry, public, and university. For example, the infrastructure planning division of the Japan Society of Civil Engineering (JSCE) established a special COVID-19 committee to conduct a consecutive survey by getting a private surveying-expert company (Survey Research Center Co., Ltd.) to cooperate nine times from April 2020 to December 2022 almost every three to five months [10]. These efforts include a fundamental report on the actual changes in activity and travel patterns due to the first pandemic, focusing on the causality between perceived risks, attitudes toward governance/self-restraint measures, and real behavioral changes [11]. Okamoto, a SRC researcher, addressed the effects of repeated pandemic waves on travel demand and mobility patterns using not only consecutive data but also other mobile spatial statistics data [12].

In addition, we can find a few detailed analyses of the influence of psychological factors such as individuals' perceived risks/attitudes toward the pandemic on actual behavioral changes in their daily lives using the survey data (see refs. [13, 14]). There are several studies on the long-term impact of the pandemic focusing on the WFH (working from home) and WFA (working from anywhere) practices. For example, in Hirose et al. [15] and Isaka et al. [16], the authors attempted to examine the behavioral characteristics of the WFH/WFA practices during the pandemic by applying their developed simulation models. The scenario-analytic approach is also applied to identify such long-term impacts in the post-pandemic period, although it is unclear how the term 'post-pandemic' would be defined in this paper [17].

COVID-19's impact on individuals' daily activity patterns can be characterized by changes in their attitudes toward the risk of the spatial and temporal spread of the pandemic waves and changes in actual behavioral patterns due to the spread of the novel coronavirus infection. Such an impact is peculiar not only to the targeted city/country but also to the point of time during the pandemic. Therefore, there is a significant difference in actual behavioral patterns, such as activity and mobility frequencies, time use, and time allocation by city/country by point of time during the pandemic. Moreover, as public preventative measures tend to be implemented in response to the extent of COVID-19 infection on a spatial scale by point of time, these governance measures would partly motivate individuals to change their attitudes/intentions toward the pandemic.

It should be noted that, in the case of Japan, they do not include such a drastic measure as the 'lockdown' applied in the US and European countries but consist of relatively moderate ones such as the government's request to restrict being out, shortened business hours, and various types of public policies, (for example, regulation, incentives, and enlightenment). A variety of changes in both attitudes toward these governance measures and actual activity and travel behaviors can be seen as short-term impacts of the pandemic in Japan. Additionally, it becomes more essential to identify how these moderate governance measures affect individuals' attitudes/

preferences, and how they would result in actual changes in their preventative behaviors (for example, self-restraint). Furthermore, it is an important challenge for us to obtain findings from the analysis on how consecutive changes in activity and travel patterns would remain in the post-pandemic era.

The emergency declaration for the first wave in Japan, in which the government requested self-restraint and promoted the introduction of telecommuting such as WFH and WFA as part of the three Cs measures, had a great effect on people's daily activity patterns, covering most trips in commuting, shopping, and eating. Particularly, the introduction of teleworking in the central core of a metropolitan area caused a decrease in the number of passengers by urban rails and in the number of visitors near terminal stations in the city center.

This effort was reported at a symposium held by the Japan Transport and Tourism Research Institute (JTTRI) in October 2020. Morichi had a keynote entitled 'The COVID impact on the demand for urban rails and the trend of urban structure," which led them to exchange their opinions at the panel discussion [18]. The agenda was to identify how we should share information on the actual situation of urban rail transport demand during the pandemic in Tokyo and discuss the impact on the urban structure in the ongoing−/post-pandemic era. Other members of the JTTRI also intended to empirically analyze the actual conditions of the number of passengers getting on and off at the gate of the targeted station using gate-tracking data during the pandemic (see refs. [19, 20]). There are several current studies on the long-term impact of commuters' teleworking on the urban structure and related changes in residential location choice behaviors [21–23]. Accessing the data on passengers passing through railway station ticket gates may not be easy; however, if possible, it would be expected to advance policy studies on office-space relocation and commercial service innovation near rail terminal stations in the post-pandemic period.

2.2 Review of COVID-19 impact studies abroad

The Organization for Economic Co-operation and Development (OECD) organized the International Transportation Forum, titled "COVID-19 TRANSPORT BRIEF," in May 2020 [24]. The forum discussed the basis of re-spacing cities for resilience: Three keywords, "react," "reboot," and "rethink" were proposed as the triple challenge that cities must meet to continue as catalysts for creating socioeconomic activity despite new health imperatives.

Palma et al. reviewed the effects of COVID-19 on mobility and lifestyle 18 months after the outbreak [25]. Short-term effects are mainly the direct ones, such as teleshopping, teleworking, air pollution, road transportation, and air travel, due to restricted mobility. On the other hand, the medium and long-term effects are regarded as indirect ones such as car ownership, employment rate, poverty, inequality, house price, relocation, and the necessity of re-designing urban spaces. They concluded that recognizing the negative effects of COVID-19 is a significant determinant of reducing the negative changes in mobility and lifestyle that affect the evaluation of individuals' psychological and social well-being.

They also discussed the challenges in simultaneously addressing the effects of COVID-19 on mobility and lifestyle. The first is the requirement of appropriate data and methodology to measure and monitor trends in changes during and after the pandemic. Second, societies would have to learn how to cope with these new structural dimensions under the remaining uncertainty, even if these changes could be measured using appropriate data and statistical modeling. During the COVID-19 pandemic, we faced not only risks but also uncertainty. The attacks of the different variants of COVID-19 showed that global changes varied to some extent under each variant. This made it difficult to assess the long-term effects. In this situation, governments are confused and cannot decide whether schools and universities should remain open and/or whether teleworking should be permanent.

Tevetkove et al. discussed the implications of the pandemic on the governance of passenger mobility innovations, such as MaaS (Mobility as a Service) in Europe [26]. They explored the challenges, barriers, and risks that the new regulatory framework must address to deploy disruptive mobility innovations. The authors' view is that, while MaaS has recently disrupted the passenger transportation sector and led requiring new and improved governance models, the COVID-19 pandemic can be seen as another disruption, stressing the need for more proactive and inclusive governance. The ongoing COVID-19 pandemic and corresponding mobility restrictions have apparently added an additional "unknown" into the already challenging mix of what the governance of disruptive mobility innovations needs to address and prioritize.

They also addressed the triggering of improved mobility service innovations by COVID-19 to progress public efficiency in the use of urban transport infrastructure, and the provision of an experimental platform, thanks to COVID-19, for the development of a new governance model essential for achieving a sustainable urban society in the future.

Hensher deliberated the role that MaaS may play post-pandemic by introducing the results of a COVID-19 impact study in Australia [27]. He aimed to explore how to reboot MaaS and rethink the overall picture of public transport by adopting the contextual approach, in which the following two scenarios were proposed: (1) business as usual, and (2) a significant change in the mobility framework. The latter is that, in a post-pandemic society, shared modes are less attractive, and WFH takes on an increasingly popular status among both employees and employers. He argued that the "new normal" offers opportunities never before achievable in terms of taming congestion on roads and crowding on public transport and that this opportunity should not be frittered away.

The primary results indicate that many sectors, such as the technology sector, have supported WFH since the pre-COVID-19 period. They also showed that WFH is expected to be a new policy lever that benefits the transport network.

These debates about the long-term impact of COVID-19 on MaaS were based on the results from the contextual approach; however, these results can indicate the preconditions with which MaaS should be equipped in response to social structure transformation, such as social distancing and WFH expansion, post-pandemic.

Mohammadi et al. focused on the long-term impacts of COVID-19 on telecommuting behaviors in the United States [28]. They implemented a comprehensive multi-wave nationwide panel survey, mainly collecting activity-based time-use data for 2020 and 2021. A panel Generalized Structural Equation Model (GSEM) was used to investigate the effects of two perceptual factors on telecommuting behavior: (1) perceived risk of COVID-19 and (2) perceived telecommuting productivity. They revealed the significant and positive impacts of productivity and COVID-risk perception on these WFH types of telecommuting behaviors. Moreover, the findings indicate that telecommuting frequency is expected to increase in the post-pandemic era, with differences across socioeconomic groups.

International collaborative studies on COVID-19 impacts supported by the World Conference on Transport Research- Special Interest Group; WCTR-SIG (organized by Zhang and Hayashi [29]) have successively produced the results of their efforts, and

Ma et al. reported their empirical analysis of the short-term impact of the first pandemic worldwide [30].

2.3 Review of COVID-19 impact studies in China

Many studies have been conducted on the effects of the COVID-19 pandemic on mobility in China. Pan et al. examined how COVID-19 changed residents' mobility using cellular signaling data from the Greater Bay Area in China [31]. The findings revealed that the COVID-19 effect on core cities was larger than that in peripheral areas and that the frequency of non-commuting travel displayed a downward tendency. Liu et al. focused on how elderly people adapted their mobility patterns and travel behaviors in Kunming, China, during the COVID-19 pandemic [32]. The results indicated that individuals older than 60 years faced challenges in riding public transport. Another study Tan et al. [33] found that Chinese mobility in 2020 was a tale of four phases, with different levels and patterns of mobility shifts. The cellular signaling dataset was used to analyze the specific properties of imparity in response and adaptation to the pandemic across cities. Zhang et al. applied a weighted stochastic block modeling approach to analyze the intercity mobility network in China in early 2020 [34]. Four types of network blocks were identified based on the mobility patterns and geographic locations of the cities. Both the COVID-19 outbreak and travel restrictions significantly disrupted the original hierarchy of the intercity mobility network in China, resulting in more local or regional fragmentation of the network, even in the recovery stage.

As for tourist mobility impact studies, Yu et al. depicted the actual changes in mobility after two waves of the COVID-19 outbreak in China using three years of cell phone data [35]. The article revealed that the number of domestic tourists in Beijing after the outbreak was higher than the pre-pandemic level, indicating a strong recovery in tourism demand. It was concluded that female and older tourists were more vulnerable to the attack in the early stages of COVID-19 but reversed their behaviors one year later, suggesting a dynamic adaptation of tourists' preferences and risk perceptions.

Zhong et al. reviewed the existing literature on tourism crisis management, focusing on the impact and implications of COVID-19 on the global tourism industry [36]. This study found that COVID-19 created opportunities for tourism stakeholders to rethink and reshape their tourism practices, such as enhancing health and safety standards, adopting digital technologies, and strengthening collaboration and coordination. Li et al. explored tourism destination management after the COVID-19 pandemic in a case study of reopened tourism destinations in China [37]. This study proposed a four-stage framework for managing tourist destinations during and after a crisis. In terms of people's willingness to travel, Hao et al. probed how COVID-19 affected people's willingness to travel for tourism based on a survey of Chinese residents in 2020 [38]. A structural equation model (SEM) was used to examine the relationships between environmental and personal factors, perceived risk and value, and willingness to travel. Perceived risk mediated the effects of environmental and personal factors on both perceived value and willingness to travel. Hong et al. applied an importance-performance analysis to evaluate the impact of bed-and-breakfasts on tourist satisfaction in Zhejiang [39].

Since the outbreak, the recovery of the tourism industry has received increasing attention. Shao et al. conducted a co-word analysis to map the topics of 140 tourism recovery policies in China [40]. Chinese tourism preferences were reflected in two rounds of nationwide online surveys in 2020 by Huang et al. [41]. The COVID-19 pandemic significantly affected tourists' travel motivations, destination choices, travel modes, information sources, and health concerns. Related implications of these shifts toward tourism recovery are also discussed. Research conducted in the literature [42] examined urban and rural tourism in China at different stages of the pandemic based on a content analysis of news reports. COVID-19 caused a sharp decline in urban and rural tourism activities but stimulated some new trends and opportunities, such as online tourism, self-driving tourism, rural revitalization, and smart tourism. They suggested that tourism stakeholders adopt a holistic, collaborative, and innovative approach to promote recovery. McCartney identified challenges and opportunities for Macao's tourism industry during and after the lockdown, such as the loss of revenue the need for innovation, and the potential for diversification [43]. Wen et al. predicted the potential effects of COVID-19 on the lifestyles and travel of Chinese citizens through a scenario analysis [44]. First, travelers' consumption patterns, such as the growing popularity of independent travel and wellness tourism, may vary. Second, new forms of tourism including slow tourism may drive future tourism activities.

3. Basic results from the 2020 and 2022 data

3.1 The outline of used 2020/2022 WEB-based data

Table 1 shows the distributions of individual attributes in the web-based datasets for 2020 and 2022. These datasets were not adaptive enough to prepare common designs and implementations in advance but resulted in the ad hoc addition of targeted cities in response to the spread of the pandemic in China. The questionnaires contained in the 2022 survey sheets were partly modified based on lessons from those of 2020, and the number of targeted cities increased from three in 2020 to seven in 2022. In addition, the web-based sampling in 2020 was different from that in 2022; significantly different patterns of sample distributions by gender, age, and whether students or not, between these two datasets, respectively, are shown in **Table 1**.

Individual attributes	2020 DATA		2022 DATA
Number of samples Targeted cities	Total: 1050 individuals [=350per city × 3 cities] Zhejiang, Beijing, Jiangsu		Total: 2450 individuals [=350per city × 7 cities] Zhejiang, Beijing, Jiangsu, Shanghai, Guangdong, Sangdong, Liaoning
Gender	male	50.3%	60.3%
	female	49.7%	39.7%
Age (6 classes)	less than 25 yrs. old	15.7%	1.8%
	25–34 yrs. old	28.1%	41.8%
	35–44 yrs. old	25.4%	33.2%
	45–54 yrs. old	30.8%	16.9%
	55–64 yrs. old	0.0%	5.0%
	over 65 yrs. old	0% (not applicable)	1.3%

Individual attributes	2020 DATA		2022 DATA
Age (3 classes)	less than 35 yrs. old	43.8%	43.6%
	35–65 yrs. old	56.2%	55.1%
	over 65 yrs. old	0.0%	1.3%
Yearly income	average (CNY/JPY)	121,800 CNY	180,600 CNY
		1,957,000 JPY	2,901,800 JPY
Household size	single-person	1.5%	1.4%
	two-persons	7.7%	13.5%
	three-persons	75.0%	65.4%
	four-persons	14.1%	15.8%
	five or more persons	1.7%	3.9%
LCS (Life Cycle Stage)	1. Younger single (younger person <50)	1.3%	1.4%
	2. Younger childless couple. (younger person <50)	5.8%	12.8%
	3. Three-generation family (youngest child <18 and living with grandparents)	7.0%	5.1%
	4. Pre-school nuclear family (youngest child <6)	18.6%	25.8%
	5. Young school nuclear family (youngest child ≥6 and < 12)	18.8%	20.9%
	6. Older school nuclear family (youngest child ≥12 and < 18)	17.1%	12.5%
	7. All adults (all ≥18)	27.9%	20.8%
	8. Single-parent and youngest child <18	0.8%	1.0%
Career (current state of individuals career, in Oct. 2020/April 2022)	company employee, government employee, and part-time job worker	79.8%	84.2%
	self-employee business	16.7%	12.6%
	Students	3.4%	0.0% (not applicable)
	others (full-time homemaker and retiree)	0.1%	3.2%

Table 1.
Sample profiles of the web-based survey datasets for 2020 and 2022.

3.2 The impact on commuters' mobility and activity patterns

We begin by introducing the trends in commuters' activities and travel behaviors by comparing the 2020 data with the 2022 data. The result from the 2020 web-based survey indicates that the average number of WFH days per week is only 0.95 days per week since the percentage without WFH (zero time per week) accounts for 57.6% and the other two types are twice a week (15.1%) and once a week (13.6%). It also shows that there was no significant difference in trends among the cities targeted in the 2020 web-based survey. On the other hand, as shown in **Table 2**, the 2022 web-based survey indicated an increasing tendency (1.39 WFH days per week). We also found that the average number of WFH days per week in Shanghai is outstanding, accounting for 3.31 days per week. It is presumed that in the case of Shanghai, because the large-scale lockdown measures were applied in response to the spread of the pandemic by the appearance of the omicron-type variation in the survey period (in April 2022), such a specific situation of the pandemic would lead to that trend in the WFH practice in the city.

Table 3 shows the distribution of the commuting modes mostly used in the latest week of the 2022 data. The results show that the ratio of cars used for commuting exceeds 60% for both employees and the self-employed. Comparing these mode-use patterns by the target city in the 2020 data with the 2022 data, they tend to depend on the level of improvement in the transportation network system in the city. Particularly, in the case of the 2022 data, the car-use ratio reached 73% in both Guangdong and Liaoning. However, in the case of Shanghai, the car-use ratio was relatively low (51.2%), but the ratio of busses, including BRT (30.1%), was much higher than that of the other modes.

Next, focusing on the changes over the years in the COVID-19 pandemic impact on commuters' behaviors, we introduce the results from comparisons of 2020 and 2019, with the ratio of 2022 to 2021 regarding the changing patterns (decrease/unchanged/increase) of frequencies of WFH, business meetings, and public transport–/car-use.

Figure 1 shows the comparisons of the ratio of 2020 to 2019, with the ratio of 2022 to 2021 regarding the changing patterns of frequencies of WFH and business meetings. In the case of WFH, both "increase" and "decrease" changing patterns exist, depending on the pandemic situation and related restriction measures in the city. This means that, in the short-term range, the percentage of "the decreasing patterns of the WFH-frequency" (from the ratio of 2020/2019 to that of 2022/2021) would be higher if the pandemic situation improved, but on the other hand, the percentage of "the increasing patterns of the WFH frequency" would be higher if the pandemic situation

No. of days working home	Beijing	Shanghai	Guangdong	Zhejiang	Jiangsu	Sangdong	Liaoning	SUM
0 day	40.2%	9.3%	58.8%	46.0%	63.4%	68.0%	56.9%	50.0%
1 day	9.5%	4.1%	11.5%	16.1%	3.6%	10.2%	10.3%	9.5%
2 days	20.8%	14.9%	13.4%	17.9%	13.4%	12.9%	15.0%	15.5%
3 days	15.5%	30.1%	7.7%	10.6%	10.1%	4.2%	9.7%	12.1%
4 days	6.0%	4.5%	2.2%	6.5%	4.5%	1.8%	3.2%	4.1%
5 days	6.3%	34.9%	3.2%	2.9%	5.1%	2.4%	4.1%	7.7%
6 days	1.8%	1.1%	3.2%	0.0%	0.0%	0.6%	0.6%	1.0%
7 days	0.0%	1.1%	0.0%	0.0%	0.0%	0.0%	0.0%	0.1%
SUM	100%	100%	100%	100%	100%	100%	100%	100%
AVERAGE	**1.63**	**3.31**	**1.05**	**1.24**	**1.04**	**0.71**	**1.07**	**1.39**

≥40%, 30%≤ <40%, 20%≤ <30%, 15%≤ <20%, 10%≤ <15%.

Table 2.
Distribution of WFH-days per week (2022 DATA).

Mostly used travel modes	Employees	Self-employed	Total
Walk(only)	0.8%	1.4%	0.9%
Bicycle	9.7%	9.0%	9.6%
Busses (BRT)	12.0%	3.5%	10.8%
Carterd busses	0.2%	0.3%	0.2%
Subways	10.0%	6.6%	9.6%
Rails	1.2%	0.0%	1.1%
Streetcars/LRT	1.0%	0.7%	1.0%
Taxi	1.4%	2.4%	1.6%
Cars	63.6%	76.1%	65.3%
Others	0.0%	0.0%	0.0%
SUM	100%	100%	100%
No. of individuals	1773	289	2062

≥15%, 10%≤ <15%, 7%≤ <10%.

Table 3.
Distribution of most-used modes for commuting (2022 DATA).

Figure 1.
The ratio of 2020/2019 and that of 2022/2021 regarding the changing patterns of frequencies of WFH and business meetings.

worsened. However, in the case of business meetings, the decreasing pattern from the ratio of 2020/2019 to that of 2022/2021 becomes higher over the years. This implies that these self-restrictive behaviors have gradually spread to working styles to reduce the risk of contagion. However, with the spread of the pandemic and related restrictive behavioral measures, many workers were deprived of going out for business meetings. In any case, it is not easy for us to reach a decision to forecast future trends in working styles, that is, how the WFH and business meeting frequencies per week would remain in China in the post-pandemic era. Therefore, it is essential to collect data on the changes in working styles in relation to the spread of the pandemic. In addition, we should turn our attention to other sociodemographic factors, such as ICT innovations and what types of industry can

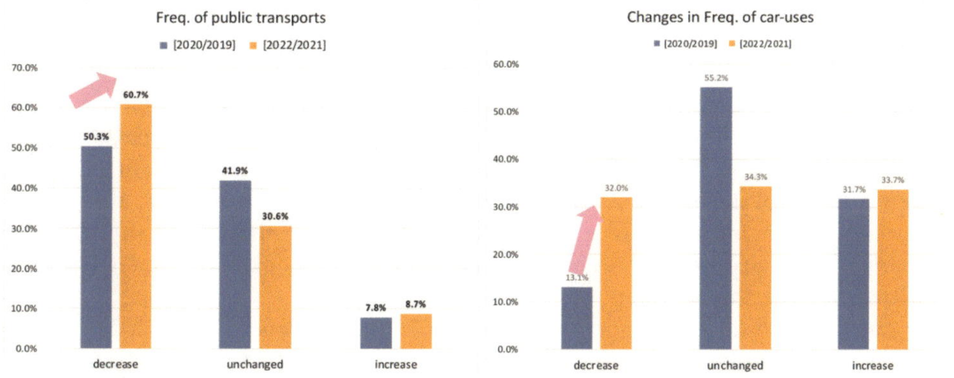

Figure 2.
The ratio of 2020/2019 and that of 2022/2021 regarding the changing patterns of frequencies of public transport- and car use in commuting.

allow high productivity of WFH, because they would clearly determine the future trend toward a post-pandemic society.

Figure 2 shows a comparison of the ratio of 2020/2019 with that of 2022/2021 regarding the changing patterns of frequencies of public transport- and car use in commuting. In the case of public transport use, we found the same changing pattern in the targeted cities in China as in European cities, as previously mentioned. People have shifted their commuting modes from public transport to their own cars. **Figure 2** also indicates that in the case of car use, the decreasing patterns from the ratio of 2020/2019 to that of 2022/2021 become higher over time. This suggests that other reasons as to why they decreased car use may exist; for example, the decrease in the total number of commuting trips by car due to the penetration of WFH over the years, and the shift from cars to other modes such as e-scooters and e-bikes.

3.3 The COVID-19 impact on daily shopping behavior

Table 4 presents a comparison of the basic characteristics of shopping behavior during the last week between two points of time: 2020 and 2022. The table indicates that the average net value of shopping behavior frequencies in 2022 increases to 2.37 in three cities, compared with the value of 1.93 in 2020. This implies that the COVID-19 impact had been slightly relieved, showing signs of returning to previous shopping patterns.

No. of days /week for shopping	3 cities: 2020 DATA	3 cities: 2022 DATA	3 cities: 2022 DATA
The percent of '0 day in a week'	8.8%	4.6%	14.4%
AVERAGE [Gross]	1.76	2.26	1.98
AVERAGE [Net]	1.93	2.37	2.32

Note: 3 cities: Beijing, Zhejiang, and Jiangsu (targeted cities in 2020 DATA).
7 cities: Shanghai, Guangdong, Shandong, and the above 3 cities.

Table 4.
Frequency of days going out for shopping in the latest week.

On the other hand, **Figures 3** and **4** show the comparisons of the ratio of 2020/ 2019 with that of 2022/2021 regarding the changing patterns of both frequencies of shopping trips per week and those of public transport- and car use, respectively. **Figure 3** indicates that in the case of shopping trips per week, the percentage of the decreasing pattern from the ratio of 2020/2019 to that of 2022/2021 increases by 22 points over the years. This may pose a question to us: How can we explain why the average number of days per week going out for shopping would increase, although the decreasing pattern increases over the years? We cannot definitely state anything at this moment, because we lack an understanding of shoppers' state dependencies under the different situations of the pandemic over the years.

Figure 4 shows a comparison of the ratio of 2020/2019 with that of 2022/2021 regarding the changing patterns of frequencies of public transport- and car use in shopping trips. **Figure 4** indicates that the result of the mode-use pattern for shopping trips has a similar tendency to that of commuting trips, as shown in **Figure 2**. When focusing on the extent to which the decreasing pattern of public transport use has changed over the years, the percentage of shopping trips increased by 17 points, but that of commuting trips increased by 10 points. This implies that the changing pattern in shopping trip frequency is more sensitive than the commuting trip frequency.

Next, we introduce the results of the COVID-19 impact on the attitude and perception of the risk of coronavirus infection in daily shopping activities and travel patterns, using the datasets of the web-based survey at two points in time. **Table 5**

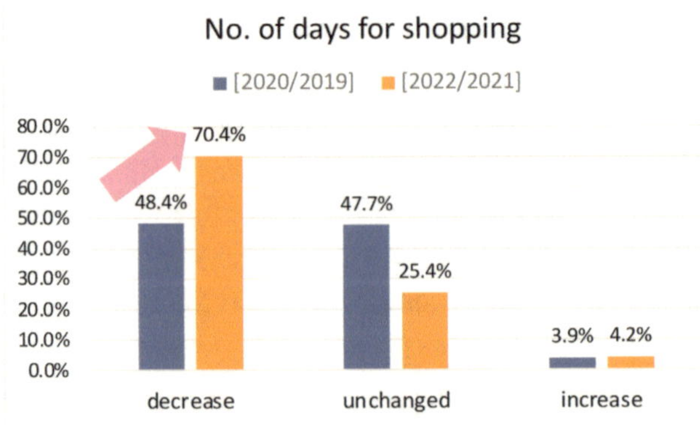

Figure 3.
The ratio of 2020/2019 and that of 2022/2021 regarding the changing patterns of frequencies of shopping trips per week.

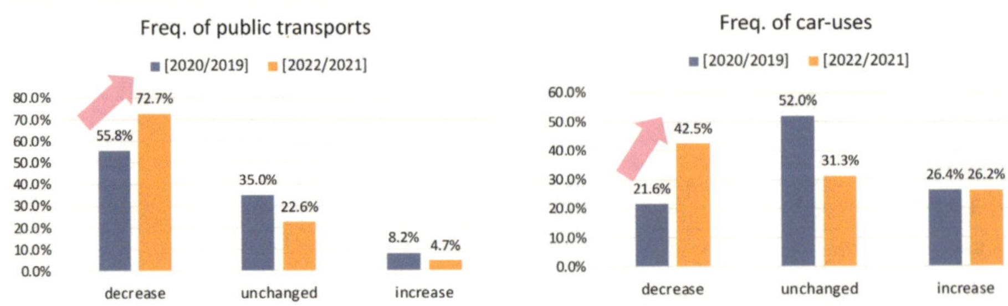

Figure 4.
The ratio of 2020/2019 and that of 2022/2021 regarding the changing patterns of frequencies of public transport- and car use in shopping trips.

Reason for 0 day for shopping	Beijing	Shanghai	Guangdong	Zhejiang	Jiangsu	Sangdong	Liaoning	SUM
abstain from going out	75.9%	69.2%	40.5%	42.9%	58.3%	42.1%	58.5%	60.2%
take delivery services	17.2%	27.3%	31.1%	42.9%	33.3%	31.6%	26.8%	28.0%
not necessary to go out	6.9%	3.5%	28.4%	14.3%	8.3%	26.3%	14.6%	11.9%
SUM	100%	100%	100%	100%	100%	100%	100%	100%

Table 5.
Distribution of the primary reasons for respondents not going shopping in the latest week (2020 DATA and 2022 DATA).

shows a comparison of the distribution of the primary reasons why each individual did not go shopping in the last week of 2020 and 2022. In the case of the 2020 data, the top-ranking reason is, "'I happened to not need to go shopping by myself or I asked my family member to do it," accounting for 42.4%. The second reason is "I use delivery services instead of going out for shopping as much as possible" (30.4%), and the third is "I abstain from going out for shopping with the intention of avoiding the COVID-19 infection risk" (27.2%). On the other hand, in the case of the 2022 data, the top-ranking reason is "I abstain from going out shopping" and the percentage jumped to 60.2%; on the contrary, the percentage of the reason "I happened to not need to go shopping by myself" dropped to 11.9%.

Our surveys asked respondents to answer the question of how they mitigated COVID-19 risks during their daily shopping activities and travel behaviors in the last week. **Figure 5** shows the distribution of behavioral attitudes toward COVID-19 in daily shopping activities and mobility, choosing the corresponding alternative from the following six options:

Alternative 1: I try to abstain from going out for shopping with the intention of avoiding COVID-19 infection risk and/or I take delivery services instead of going out for shopping as much as possible.

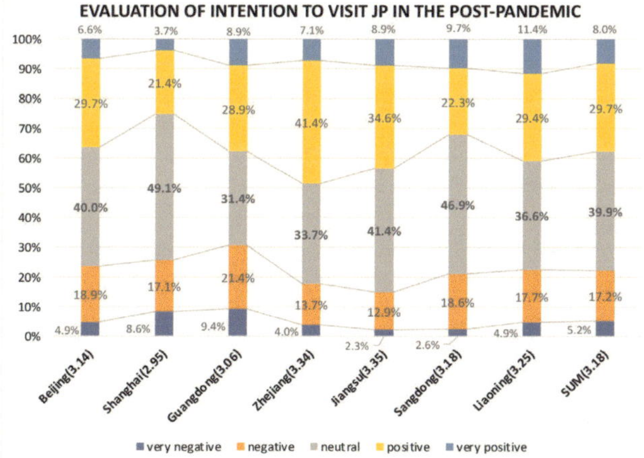

Figure 5.
Distribution of the intention of visiting Japan and the average of evaluated scores under the improved situation of the COVID-19 pandemic by targeted city (2022 DATA).

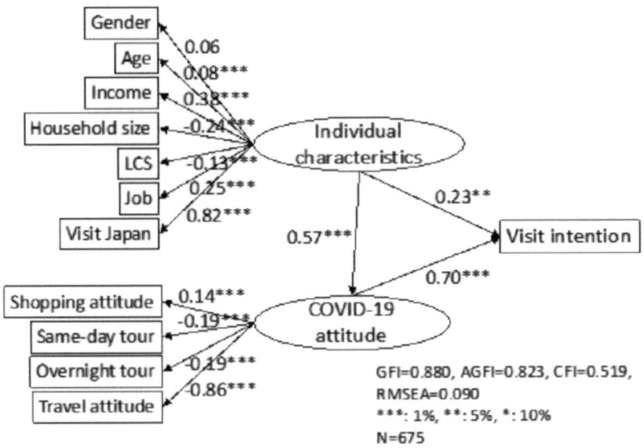

Figure 6.
Result of the standardized parameter estimates of the SEM-model (in case of the 2020 DATA).

Alternative 2: I try to avoid using congested public transport (busses and subways) to access shopping places.

Alternative 3: I avoid shopping at congested supermarkets and shops.

Alternative 4: I try to reduce the duration of staying in shopping places.

Alternative 5: I try to maintain social distancing between people and wear a mask.

Alternative 6: Others (other reactions/behavioral attitudes toward COVID-19 infection risk).

Figure 6 indicates that these behavioral attitudes in both the 2020 and 2022 data were distributed equally, except for the sixth alternative (others). The percentage of Alternative 5, "I try to keep social distancing between persons and wear a mask," increases from 12.5% in 2020 to 25.8% in 2022, and the percentage of Alternative 1, "I try to abstain from going out for shopping," does from 13.1% to 17.9% as well. This implies that attitudes toward COVID-19 infection risk in relation to daily shopping behaviors vary widely; however, it takes a long time for individuals to respond to the spread of the pandemic and related governance measures in China (**Table 6**).

How to care for COVID-19?	2020 DATA	2022 DATA
abstain from going out shopping	13.1%	17.9%
avoid using public transit	22.5%	18.0%
avoid the congested	26.9%	19.2%
diminish the duration	25.0%	19.1%
put on mask	12.5%	25.8%
other reactions of COVID-19	0.0%	0.0%
SUM	100%	100%

Table 6.
Distribution of behavioral attitudes toward COVID-19 in daily shopping activity and related mobility (2020 DATA and 2022 DATA) (multiple answers).

4. Discussion: the long-term impact of COVID-19 on the intention of visiting Japan

4.1 The pandemic risk and intention of overseas travel (including visit-to-Japan tours)

Our web-based surveys in 2020 and 2022 matched the timing of implementation with the spread of the pandemic and corresponding governance measures in China, respectively. The 2020 web-based survey was conducted in November 2020. This timing was chosen because the government declared a victory over the first wave of the COVID-19 pandemic at the National People's Congress in May 2020 and celebrated the containment of the pandemic in July 2020 through governance measures (such as gate-checks by introducing health care app and an automatic tracking system) were successively led to relieve behavioral restrictions in the targeted cities of China until the end of October 2020.

However, when we started to design questionnaires for the 2022 web-based survey, they experienced the rapid spread of COVID-19 owing to the appearance of new types of coronaviruses in China. Under the circumstances of the return of the pandemic to China, the Chinese government had been maintaining the zero-coronavirus measure, that is, the perfect lockdown in the infected districts/cities, including traffic restriction/prohibition and quarantine/isolation rules. As a result, many cities in China are diversifying both the inhabitants' attitudes toward infection risks and actual behavioral changes.

In the 2022 WEB-based survey, respondents answered questions about the perceived level of COVID-19 infection in their residential areas. The perceived level was classified into seven categories: Level 1 (no restriction but social distancing) to Level 7 (lockdown in the whole city). **Table 7** shows the distribution of the levels of respondents' perceptions of the spread of COVID-19 infection in the whole of 7 targeted cities. **Table 8** presents the city distributions.

Table 7 indicates that the percentage of relatively low levels of perceived infection (from Level 1 to Level 3) only accounts for 50%, and the remaining half of the respondents perceived more severe levels of COVID-19 infection in their residential districts/cities. **Table 8** indicates that in Shanghai, the percentage of such low levels of infection is only 25%, but that of the middle levels (levels 4 and 5) accounts for

Level (4 classes)	Level (7 classes)	Level of restriction for COVID-19 in surveyed point in time {2022DATA]	SUM
Lowest	Level 1	no restriction but social distances	3.4%
Low	Level 2	health-coding identification required	27.0%
	Level 3	restrictive to use facilities	19.2%
Middle	Level 4	promoting remote-work and stay-home	29.2%
	Level 5	restriction on the movement (permission required)	13.2%
High	Level 6	prohibition on the movement in a specific district	4.8%
	Level 7	lock-down in the whole city	3.1%
		SUM	100%

Table 7.
Distribution of levels of behavioral restrictions as an indicator of the perceived COVID-19 infection.

	Level	Beijing	Shanghai	Guangdong	Zhejiang	Jiangsu	Sangdong	Liaoning
1	Level 1	5.4%	2.3%	3.4%	5.7%	2.9%	3.1%	1.1%
2	Level 2	25.7%	12.9%	27.4%	38.6%	19.7%	39.7%	25.1%
3	Level 3	15.1%	12.0%	26.9%	15.1%	22.6%	22.0%	20.9%
4	Level 4	34.0%	44.3%	33.7%	22.0%	26.9%	19.7%	23.7%
5	Level 5	11.7%	14.6%	6.6%	10.6%	18.9%	10.3%	20.0%
6	Level 6	5.4%	7.4%	1.7%	4.0%	5.7%	2.9%	6.6%
7	Level 7	2.6%	6.6%	0.3%	4.0%	3.4%	2.3%	2.6%
		100%	100%	100%	100%	100%	100%	100%
	average	3.47	4.05	3.19	3.21	3.70	3.12	3.66

≥40%, 30%≤ <40%, 20%≤ <30%, 10%≤ <20%, 5%≤ <10%.

Table 8.
Distribution of levels of behavioral restrictions as an indicator of the perceived COVID-19 infection by targeted city.

approximately 60%. The table also indicates that the perceived levels of infection were widely distributed in other targeted cities.

In China, the government tends to invoke strong-arm political tactics to achieve complete prevention against the pandemic, but at the same time, the people have often suffered from lack of information about the actual spread of the pandemic and a tight situation for beds in hospitals. Such problems, although specific to China, maybe one of the causes of diversification in the perception of the COVID-19 pandemic and attitudes toward infection risk.

Next, we introduce results on how Chinese individuals intend to travel overseas during the current situation of COVID-19, that is, at the surveyed point in time (November 2020 in the case of the 2020 data and March 2022 in the case of the 2022 data). The question was 'How do you think your overseas travel is in the current situation?' A single answer was selected from the following five alternatives:

Alternative 1: '1. cannot think of my overseas travel for a while.'
Alternative 2: '2. it is difficult to think about my overseas travel'.
Alternative 3: '3. unwilling to think my overseas travel'.
Alternative 4: '4. would reduce overseas travel opportunities'.
Alternative 5: '5. would like to positively think my overseas travel'.

How do you think your overseas travel under the current situation of COVID-19?	2020 DATA	2022 DATA
(1) can not think about my overseas travel	18.6%	18.3%
(2) difficult to think about my overseas travel	37.7%	32.2%
(3) not willing to think about my overseas travel	10.2%	9.2%
(4) reduce overseas travel opportunity	10.9%	14.2%
(5) would like to enjoy my visit to Japan	22.7%	26.1%
SUM	100%	100%

Table 9.
Distribution of the intention of overseas travel during the current situation of the pandemic (2020 DATA and 2022 DATA).

How do you think overseas travel?	Beijing	Shanghai	Guangdong	Zhejiang	Jiangsu	Sangdong	Liaoning
(1) can not think overseas travel	17.4%	20.3%	13.7%	15.7%	18.3%	23.7%	19.1%
(2) difficult to think overseas travel	28.0%	32.3%	32.9%	38.9%	28.0%	37.7%	27.7%
(3) unwilling to think overseas travel	10.9%	9.4%	15.4%	8.6%	6.0%	5.7%	8.3%
(4) reduce overseas travel opportunity	21.4%	12.3%	15.7%	15.4%	12.3%	10.9%	11.4%
(5) would like to enjoy visiting Japan	22.3%	25.7%	22.3%	21.4%	35.4%	22.0%	33.4%
	100%	100%	100%	100%	100%	100%	100%

≥35%, 30%≤ <35%, 25%≤ <30%, 20%≤ <25%, 15%≤ <20%, 10%≤ <15%, 5%≤ <10%.

Table 10.
Distribution of the intention of overseas travel during the current situation of the pandemic by targeted city (2022 DATA).

Table 9 shows distributions of the intention of overseas travel under the current situation of the pandemic at two points of time; 2020 data and 2022 data. **Table 10** indicates the result of such distribution by city in the 2022 data. **Table 9** results that those intentions of overseas travel in the current situation in China vary widely between alternatives and that there is little difference in distribution and ordering between the two points of time.

It is however noted that when our dividing such distribution of the intentions into two parts; 'negative' (Alternative 1 and 2) and 'positive' (Alternative 3, 4, and 5), the percentage of 'negative' versus the one of 'positive' in 2020 is 56.3% versus 44.7%, but on the other hand, 50.5% versus 49.5% in 2022 and that the percentage of 'positive' in 2022 increases with 5 points comparing with that in 2020. **Table 10** also indicates that in the case of the 2022 data, the distributions of such intentions by target city differ significantly from each other. These results may partly reflect the heterogeneity of individuals' behavioral attitudes toward COVID-19, assuming that such heterogeneity could be created through various changes in socioeconomic circumstances due to COVID-19 infection in the city.

In both our web-based surveys in 2020 and 2022, we asked a question about the intention to visit Japan under the precondition that the COVID-19 situation would improve and overseas travel restrictions would be lifted. Each respondent answered this question on a five-point Likert scale ranging from 1 (very negative) to 5 (very positive).

Let us denote three groups; "negative group," "neutral group," and "positive group." The negative group is here defined as those who select "1. very negative" and "2. negative." On the other hand, the positive group contains those who select "4. Positive" and "5. very positive" and the neutral group is equal to "3. neutral."

Table 11 shows a comparison of the distribution of the intention to visit Japan and the average of the evaluated scores under the improved situation between the 2020 and 2022 data. **Figure 5** shows the distribution of these intentions and the average for the targeted city. Each respondent answered the question under the hypothesized situation of the pandemic, that is, under the condition that the situation surrounding the COVID-19

Evaluation of intention to visit JP	2020 DATA	2022 DATA
1 very negative	6.4%	5.2%
2 negative	13.0%	17.2%
3 neutral	27.8%	39.9%
4 positive	40.9%	29.7%
5 very positive	12.0%	8.0%
SUM	100%	100%
AVERAGE	3.39	3.18

Table 11.
Distribution of the intention of visiting Japan and the average of evaluated scores under the improved situation of the COVID-19 pandemic (2020 DATA and 2022 DATA).

pandemic could be improved. The results indicate that the percentage of positive intention to visit Japan in the 2022 data decreases by 15.2 points, but that of neutral intention increases by 12.1 points. This implies that such negative changes in attitudes toward overseas travel occurred partly because of the prolonged impact of the pandemic.

Figure 5 indicates that, in the case of Shanghai, the average evaluated score of intention to visit Japan under the improved situation is 2.95 in 2022 data and it was the lowest among all the targeted cities. We may say that the seriousness of the pandemic in many districts of Shanghai in those days may be one of the causes of such a low intention to visit Japan, even in an improved situation. On the other hand, in the case of Zhejiang and Jiangsu, the average scores are highly regarded; 3.34 in Zhejiang and 3.35 in Jiangsu. The results suggest that the actual situation of the pandemic had relatively gone into remission in those days, and the attitudinal factors regarding the risk of uncertainty of COVID-19 could have restrained its performance. In addition, the negative factors determining the intention to visit Japan under the improved situation were also controlled.

Selecting the subsamples from our pooled 2020 and 2022 datasets, the SEM-model (SEM; Structural Equation Modeling) is here applied to identifying the causal structure underlying the intention of visiting Japan in anticipation of the post-pandemic. SEM studies have recently spread the range of fields applied not only to travel behavioral data but also to tourism marketing data. This is partly because statistical software packages, such as IBM SPSS Amos [45], have become more popular in these fields [46–47]. Kurihara et al. also developed an SEM model to explore the causal relationship between integrated mobility services and the intention to revisit Japan using 2019 interview survey data from Chinese tourists who visited Japan and left Osaka International Airport (KIX) for China [48]. Basically, the causal structure of our developed SEM model is similar to that of the previous model, which is featured with simplification for the purpose of verifying the basic causality between the COVID-19 impact and the intention of visiting Japan in the post-pandemic period.

In our developed SEM model, there are two latent variables: One is called "Individual characteristics" and the other is "COVID-19 attitudes." These two variables play the role of latent factors that represent causality among the observed variables, such as the intention to visit Japan, the attitude and perception of the COVID-19 risk, and the experience of visiting Japan, including individual attributes. **Table 12** presents the definitions of the variables used in this study.

Figures 6 and 7 show the diagrams representing causal paths between "Intention of visiting Japan" and two latent variables revealing the estimation result of

Variable	Definition
Gender	1 = male, 0 = female
Age	1 = 40, 50, 60's, 0 = 20, 30's
Income	1 = 1, 2 = 3,3 = 5, 4 = 7,5 = 9,6 = 12.5, 7 = 17.5,8 = 22.5,9 = 27.5,10 = 60 (thousand yuan)
Household size	1 = single, 2 = two, 3 = three, 4 = four, 5 = more than five
LCS	1 = family with children, 0 = others
Job	1 = self-employee, 0 = others
Shopping attitude	1 = abstain from going out for shopping, 0 = others
Same-day tour	1 = decreased, 0 = other
Overnight tour	1 = decreased, 0 = other
Visit intention	1 = very negative to 5 = very positive
Travel attitude	1 = abstain from travel abroad, 0 = others
Visit Japan	1 = have been to Japan, 0 = have not been to Japan

Table 12.
Definition of used variables in the SEM model.

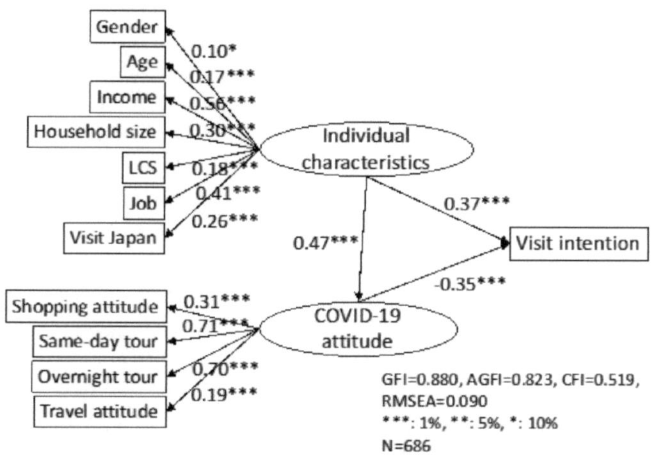

Figure 7.
Result of the standardized parameter estimates of the SEM-model (in case of the 2022 DATA).

standardized parameters of causal paths in the SEM- model in the case of the 2020/2022 pooled data set in which the individuals who live in Beijing and Zhejiang are targeted. Note that all observed variables and estimated parameters of causal paths are distinguished between two points in time, although the values of the indicators of goodness of fit, such as GFI, AGFI, CFI, and RMSEA, are the same in both figures.

The following are the results from the parameter estimates.

- The goodness of fit indicators for the model (GFI = 0.880, AGFI = 0.823, CFI = 0.519, RMSEA = 0.093) met the standard level under the assumption of a simple model structure. Almost of the null hypotheses are rejected with a 1% significant level except for the parameter estimate of 'Gender.'

- When focusing on the causal path from COVID-19 attitudes to intention to visit Japan, the sign of the parameter estimate in 2020 is positive, but that in 2022 is negative. It is however noted that the sign of the causal paths related to the observed variables except "shopping attitude" are reversed between two points of time. This implies that the direct effect of 'COVID-19 attitude' on 'Visit intention' tend to weaken over the years.

- Also, the values of the indirect effect of 'Travel attitude' on 'Visit intention' are −0.602 in 2020 and − 0.067 in 2022. This result suggests that the risk of overseas travel tends to weaken rapidly over time.

- On the other hand, the values of indirect effect of 'Overnight tour' on 'Visit intention' are −0,133 in 2020 and − 0.245 in 2022. This means that those who had a tendency to abstain from overnight tours during the pandemic would intend not to visit Japan, even in the post-pandemic era.

4.2 Perspective and challenges of inbound tourism demand in the post-pandemic era

Our analytical hypothesis is that the intention to visit Japan in the post-pandemic era could be formed as an extension of the changes in both the behavioral attitude and perception toward the COVID-19 infection risk and substantive mobility and lifestyle during these three years, that is, during the pandemic period. This study aimed to identify the challenges in verifying the hypothesized causal structures underlying the intention to visit Japan during the post-pandemic era.

Here, we discuss inbound tourism demand from China during the post-pandemic era. Many tourism marketing researchers, including airlines, travel agencies, and tourism-related industries, have vigorously discussed the impact of COVID-19 on international tourism demand from the perspective of "disruptive innovation in the economy." However, we are aware that most of their discussions started and ended with the COVID-19 impact on manageable issues, such as the decreased level of services with a reduction in the number of flights, cutbacks of airline workers, and serious financial troubles in hotels and tourism retailing and services industries. Many scenario analyses for predicting the future trend of tourism demand from the marketer-oriented approach tend to lead to a certain type of optimistic prospect that international tourism demand will come around soon if only the level of services on the supply side, such as the number of carriers, returns to normal. Focusing on the long-term impact on the inbound tourism market in the post-pandemic era, it is essential to examine how factors relating to both the marketers' supply side and tourists' demand side would determine the tourism market rebuilt in a post-pandemic society. Moreover, the supply-side-oriented approach would lead us to misunderstand our perspective of future demand if changes in lifestyles and values are assumed to be significant demand-side factors under the disruptive market mechanism.

While we reviewed COVID-19 impact studies in European cities, the basis of their tone was that COVID-19 has led to an innovative urban transport system such as MaaS to the newly disruptive innovation (see [26]). Such a tone seems to reflect a sense of crisis and may be exaggerated too much, but their raised challenges regarding the long-term impact of the COVID-19 pandemic should be shared with our discussion from the perspective of inbound tourism demand from China. One of the most important challenges is the successive collection of longitudinal tracking data to monitor dynamic changes in individual activity and mobility patterns during repeated waves of the pandemic. These longitudinal tracking data, including repeated cross-sectional survey data, are useful for adopting an activity and travel behavior approach. In particular, they are expected to contribute to an accurate representation of the causal structure underlying tourists' needs to travel overseas, as well as their intention to visit Japan.

On the other hand, several macroscopic econometric modeling approaches, such as System Dynamics and advanced urban simulation techniques, including the Machine Learning (ML) model and AI-based models, have already been applied to the scenario analysis of the future trend of travel demand post-pandemic. As another promising quantitative approach, we would like to add the model, which we have applied to forecast the long-term trend of the Tourism Area Life Cycle (denoted TALC) in targeted tourism destinations. Butler (1980) developed a prototype describing the evolution of tourism destinations. Nishii et al. developed the TALC model to analyze the trend in the number of tourists visiting Hokkaido in Japan, focusing on the effect of external factors determining the carrying capacity, such as natural disasters, infrastructure projects for improving the level of transport services in airlines, expressways, Hokkaido Shinkansen (Japan's Bullet Train), and events and campaign performance for promoting tourism marketing in Hokkaido [49]. The TALC modeling approach makes the most of the benefits of incorporating a variety of factors relating to both supply and demand into the relatively simplified time-series model structure.

Author details

Takeshi Kurihara[1*], Kazuo Nishii[2†], Yilin Sun[3] and Yinan Dong[3]

1 Toyo University, Tokyo, Japan

2 Yamanashi University, Kofu, Japan

3 Zhejiang University, Hangzhou, China

*Address all correspondence to: kurihara039@toyo.jp

† These authors contributed equally.

References

[1] NHK-Internet News-Special: COVID-19, News on Nov. 16, 2022. Beyond one million new-cases per day, need to strengthen medical services toward the eighth wave of the pandemic. Available from: https://www3.nhk.or.jp.news.special.coronavirus.eightwave [Accessed: 31 March 2023]

[2] Nishii K. Perspective of our research topics in the post pandemic era, a note presented at International Session in the 40th Anniversary KST Conference Joined up to the 11th International Seminar on Urban Transport, Tourism and Travel Behavior Analysis, Seoul. [Accessed: 28 September 2022]

[3] ERSA Web conference 2020: Spatial Challenges for the New World. Available from: https://ersa.eventstar.com [Accessed: 1 August 2020]

[4] TRB 102nd Annual Meeting 2023. https://www.trb.org.AnnualMeeting.program [Accessed: 1 April 2023]

[5] Nishii K, Kurihara T, Hibino N, Kishino K. A structural equation modeling of the intention of revisits by Chinese tourists to Japan conditioned upon an integration of mobility services in urban tourism destination: Using an empirical data set from the interview survey at Osaka International Airport, Report on 2019 Research grant by Kansai Airport Research Institute (KAR), 2020.3 (in Japanese/English)

[6] Nishii K, Kurihara T, Hibino N, Kishino K. Mobility services integrations in tourism destination and their effect on intention of visit to Japan: Using the 2020 Web-based survey data targeted at Chinese residents, Report on 2020 Research Grant by Kansai Airport Research Institute (KAR), 2021.3 (58p. in Japanese version/68p. in English version.), 2021.3

[7] Nishii K, Kurihara T, Hibino N, Kishino K. A Basic analysis of Chinese tourists' intention of visiting Japan relating to attitudinal changes in their activity and travel patterns under the with–/post-COVID-19: Using the WEB-based survey data in three major cities (Zhejiang, Beijing, and Jiangsu), Proceedings of the JSCE Infrastructural Planning and Transportation Division, No. 63, CD-ROM, pp. 1-7, 2021. 6 (in Japanese)

[8] Kurihara T, Nishii K, Hibino N, Kishino K. The impact of preference for integrated mobility services on visit intention-A case study of Chinese traveling to Japan, presented at International Session in the 40th Anniversary KST Conference, 2022. 9

[9] Ministry of Land, Infrastructure, Transport and Tourism (denoted MLIT): Discussion on how changes in the future urban renovation and compact cities would be triggered by the COVID-19, Press release, 21p., 2020.8

[10] The JSCE Committee of Infrastructural Planning and Transportation Division: The survey of the impact of the COVID-19 on mobility and attitudes (instantly reported version); adding the 9th survey data. Available from: https://jsce-ip.org.2022/09/16.COVID-19-survey, 2023.1(in Japanese) [Accessed: 24 May 2023]

[11] Hyodo T, Fujii S. Recent advances on the COVID-19 related studies in the academic field of the infrastructural planning and transportation. Proceedings of the JSCE, D3-division. Vol. 77(2):110-112, 2021.1 (in Japanese)

[12] Okamoto T. The changes in urban travel demand caused by the COVID-19 with using the actual surveyed data. Journal of Transportation Sciences. 2021; **52**(2):13-20 (in Japanese)

[13] Takeda M, Ishibashi S, Taniguchi M. Behavior elasticity chart showing behavioral changes caused by the COVID-19 pandemic: Quantification of return and non-return to pre-pandemic state, Proceedings of the JSCE, D3-division, Vol. 78, No.6, II_95-II_107, 2022.5 (in Japanese)

[14] Ishibashi T, Taniguchi A, Parady G Takami K. Time-series analysis of behavioral changes and psychological state of citizens in the Kanto region during the COVID-19 pandemic: Using a five-wave Panel survey from April 2020 to July 2021, Proceedings of the JSCE Infrastructural Planning and Transportation Division, No. 64, CD-ROM, pp1-8, 2021. 12 (in Japanese)

[15] Hirose K, Sasaki K, Kikuchi A, Onoda K, Asano R, Suzuki T. Activity simulation under COVID-19 situation considering the spread of remote work, Proceedings of the JSCE Infrastructural Planning and Transportation Division, No. 63, CD-ROM, 1-7, 2021. 6 (in Japanese)

[16] Isaka R, Sasaki K, Kikuchi A. Research on reproduction of individual behavior using activity simulators and observed data: Recreation of telecommuting for the corona disaster, Proceedings of the JSCE Infrastructural Planning and Transportation Division, No. 65, CD-ROM, 1-7, 2022. 6 (in Japanese)

[17] Muto T, Kaneko Y. Analysis of COVID-19 expansion and telework implementation using mobile spatial statistics in the Tokyo metropolitan area, Proceedings of the JSCE Infrastructural Planning and Transportation Division, No. 64, CD-ROM, 1-6, 2021. 12 (in Japanese)

[18] JTTRI (Japan Transport and Tourism Research Institute): Symposium on Effect of the COVID-19 on Rail Transportation and Urban Infrastructure, 2020.10.26

[19] Suzuki W, Nakamura F, Tanaka S Ariyoshi A, Matsuyuki M. Aggregation analysis of changes in habitual use patterns of railways using the data from passengers counted at the ticket gate, Proceedings of the JSCE Infrastructural Planning and Transportation Division, No. 66, CD-ROM, 1-8, 2022. 11 (in Japanese)

[20] Hashimoto M, Hibino N, Morichi S. The actual conditions of changes in railway commuting behaviors using the data from passengers counted at the ticket gate, Proceedings of the JSCE Infrastructural Planning and Transportation Division, No. 66, CD-ROM, 1-8, 2022. 11 (in Japanese)

[21] Muroi T, Tsuchiya T, Itoh M, Morichi S. A study on a location choice model focusing on the residential area along a railway line in Tokyo metropolitan areas, Proceedings of the JSCE Infrastructural Planning and Transportation Division, No. 66, CD-ROM, 1-8, 2022. 11 (in Japanese)

[22] Takeda M, Kobayashi T, Taniguchi M. The actual conditions of blunt distribution typed urban time-spaces: Focusing on time-allocation and residential area choice behavior in daily activity patterns, Proceedings of the JSCE Infrastructural Planning and Transportation Division, No. 66, CD-ROM, 1-7, 2022. 6 (in Japanese)

[23] Miura U, Mimaki H, Nakamura F, Kitazaki M, Oumori H, Yukawa S. The

future vision of Tokyo downtown based on possibility of transformation of working style: Challenges about working-styles related to the New Normal in the post-pandemic, Proceedings of the JSCE Infrastructural Planning and Transportation Division, No. 66, CD-ROM, 1-7, 2022. 11 (in Japanese)

[24] OECD: COVID-19 TRANSPORT BREIEF; Re-spacing our cities for resilience, International Transport Forum, 2020, Available from: https://www.itf-oecd.org.covid-19 [Accessed: 1 May 2022]

[25] de Palma A, Vosough S, Liao F. An overview of effects of COVID-19 on mobility and lifestyle: 18 months since the outbreak. Transportation Research Part A. 2022;**159**:372-397

[26] Tsvetkova A, Kulkov I, Busquet C, Kao PJ, Kamargianni M. Implications of COVID-19 pandemic on the governance of passenger mobility innovations in Europe. Transportation Research Interdisciplinary Perspectives. 2022;**14**: 100581

[27] Hensher DA. What might COVID-19 mean for mobility as a service (MaaS)? Transport Reviews. 2020;**40**(5): 551-556

[28] Mohammadi M, Rahimi E, Davatgari A, Javadinasr M, Mohammadian A, Bhagat-Conway MW, et al. Examining the persistence of telecommuting after the COVID-19 pandemic. Transportation Letters. 2022: 1-14. DOI: 10.1080/19427867.2022.2077582

[29] Zhang J, Hayashi Y. Transportation Amid Pandemics - Lessons Learned from COVID-19, World Conference on Transport Research Society. 1st ed. Elsevier; 2023. 490 p. eBook ISBN: 9780323997713

[30] Ma S, Li S, Zhang J. Diverse and nonlinear influences of built environment factors on COVID-19 spread across townships in China at its initial stage. Scientific Reports (Springer Nature Publishing Group). 2021. DOI: 10.1038/s41598-021-91849-1

[31] Pan Y, He SY. Analyzing COVID-19's impact on the travel mobility of various social groups in China's Greater Bay Area via mobile phone big data. Transportation Research Part A: Policy and Practice. 2022;**159**:263-281

[32] Liu Q, Liu Y, Zhang C, An Z, Zhao P. Elderly mobility during the COVID-19 pandemic: A qualitative exploration in Kunming, China. Journal of Transport Geography. 2021;**96**:103176

[33] Tan S, Lai S, Fang F, Cao Z, Sai B, Song B, et al. Mobility in China, 2020: A tale of four phases. National Science Review. 2021;**8**(11):nwab148

[34] Zhang W, Gong Z, Niu C, Zhao P, Ma Q, Zhao P. Structural changes in intercity mobility networks of China during the COVID-19 outbreak: A weighted stochastic block modeling analysis. Computers, Environment and Urban Systems. 2022;**96**:101846

[35] Yu L, Zhao P, Tang J, Pang L. Changes in tourist mobility after COVID-19 outbreaks. Annals of Tourism Research. 2023;**98**:103522

[36] Zhong L, Sun S, Law R, Li X. Tourism crisis management: Evidence from COVID-19. Current Issues in Tourism. 2021;**24**(19):2671-2682

[37] Li Z, Zhao Q, Huo T, Shao Y, Hu Z. COVID-19: Management focus of reopened tourist destinations.

Current Issues in Tourism. 2022;**25**(1): 14-20

[38] Hao Y, Bai H, Sun S. How does COVID-19 affect tourism in terms of people's willingness to travel? Empirical evidence from China. Tourism Review. 2021;**76**(4):892-909

[39] Hong Y, Cai G, Mo Z, Gao W, Xu L, Jiang Y, et al. The impact of COVID-19 on tourist satisfaction with B&B in Zhejiang, China: An importance–performance analysis. International Journal of Environmental Research and Public Health. 2020;**17**(10):3747

[40] Shao Y, Hu Z, Luo M, Huo T, Zhao Q. What is the policy focus for tourism recovery after the outbreak of COVID-19? A co-word analysis. Current Issues in Tourism. 2021;**24**(7): 899-904

[41] Huang SS, Shao Y, Zeng Y, Liu X, Li Z. Impacts of COVID-19 on Chinese nationals' tourism preferences. Tourism Management Perspectives. 2021;**40**: 100895

[42] Li Z, Zhang X, Yang K, Singer R, Cui R. Urban and rural tourism under COVID-19 in China: Research on the recovery measures and tourism development. Tourism Review. 2021; **76**(4):718-736

[43] McCartney G. The impact of the coronavirus outbreak on Macao. From tourism lockdown to tourism recovery. Current Issues in Tourism. 2021;**24**(19): 2683-2692

[44] Wen J, Kozak M, Yang S, Liu F. COVID-19: Potential effects on Chinese citizens' lifestyle and travel. Tourism Review. 2021;**76**(1):74-87

[45] IBM SPSS Amos. Available from: http.www.ibm.com/jp-ja/marketplace/structural-equation-modeling-sem [Accessed: 11 May, 2023]

[46] Sasaki K. In: Kitamura R, Morikawa T, editors. Modeling Travel Behavior, Gihodo-Shuppan Publishing, the Third Part: Modeling, in Chapter 8. Gihodo-Shuppan Publishing Co Ltd.; 2002. pp. 159-176

[47] Nishii K, Sasaki K, Tanabe J. Empirical analysis of leisure driver decision structure underlying SA/PA choice behaviors on expressways in Japan. In: Paper Presented at the Ordinary Oral Session in the 53rd ERSA Congress in Palermo. Italy; 2013

[48] Kurihara T, Nishii K, Hibino N, Kishino K. A structural equation modelling (SEM) of the causal relationships between mobility services integration in tourism destinations and the intention of revisits to Japan by Chinese tourists, Proceedings of the JSCE, Special issue-D, 76/5, I_747-I_756, 2021 (in Japanese)

[49] Nishii K, Morisawa T, Mitsuishi T: A Quantitative Analysis of Tourism Area Life Cycle (TALC): From viewpoint of the AMMA (Area Marketing and Management Approach), Proceedings of the 34th JITR (Japan Institute Tourism Research) Annual Conference, 1-4, CD-ROM, 2019.12 (in Japanese)